NUCLEAR MEDICINE IN GASTROENTEROLOGY

Developments in Nuclear Medicine

VOLUME 18

Series Editor: Peter H. Cox

Consulting Editor: Henry N. Wagner

The titles published in this series are listed at the end of this volume.

NUCLEAR MEDICINE IN GASTROENTEROLOGY

edited by

Hans J. Biersack
Department of Nuclear Medicine, University of Bonn,
Bonn, Germany

and

Peter H. Cox
Department of Nuclear Medicine, Dr. Daniel den Hoed Clinic,
Rotterdam, The Netherlands

SPRINGER SCIENCE+BUSINESS MEDIA, B.V.

Library of Congress Cataloging-in-Publication Data

Nuclear medicine in gastroenterology / edited by Hans J. Biersack and
Peter H. Cox.
p. cm. -- (Developments in nuclear medicine ; v. 18)
Includes index.
ISBN 978-94-010-5464-5 ISBN 978-94-011-3316-6 (eBook)
DOI 10.1007/978-94-011-3316-6
1. Digestive organs--Radionuclide imaging. I. Biersack, H. J.
II. Cox, Peter H. III. Series: Developments in nuclear medicine ;
18.
[DNLM: 1. Diagnostic Imaging. 2. Gastrointestinal Diseases-
-radionuclide imaging. 3. Nuclear Medicine. W1 DE998KF v. 18 / WI
141 N9645]
RC804.R27N835 1991
616.3'307575--dc20
DNLM/DLC
for Library of Congress 90-15636

ISBN 978-94-010-5464-5

Printed on acid-free paper

Table of contents

Preface

During the last two decades significant advances have been made in the in vivo-diagnosis of gastrointestinal diseases. Although Ultrasound and CT as well as Endoscopy have had a major impact on the evaluation of liver, pancreas and bile diseases, there are a lot of indications for Nuclear Medicine procedures. These include new investigational procedures like esophageal scintigraphy, proof of bleeding sites, scintigraphy of inflammatory diseases, and intestinal resorption tests. Further, immunoscintigraphy with radiolabelled antibodies has gained wide-spread application especially in colon cancer. The differential diagnosis of liver tumors like haemangioma and focal nodular hyperplasia by means of blood pool and HIDA-scintigraphy is nowadays a routine procedure. Other established methods like hepatobiliary scintigraphy and liver perfusion scintigraphy have proved to be reliable tools in the pre- and postoperative evaluation of patients with bile duct obstruction and portal hypertension. The aim of this book is to present the entire spectrum of Nuclear Medicine in Gastroenterology to our colleagues from internal medicine and surgery. Ultrasound and Sonography as well as CT will rule the field of gastroenterology, but there remain a certain number of unanswered questions. Nuclear Medicine provides a lot of reliable answers.

H.J. Biersack and P.H. Cox
July 1990

List of contributors

Duncan M. Ackery, Department of Nuclear Medicine, Southampton General Hospital, Tremona Road, Southampton, Hampshire SO9 4XY, U.K.

Roland Bares (co-author: U. Buell), Department of Nuclear Medicine, Technical University of Aachen, Pauwelsstr. 1, DW-5100 Aachen, Germany.

Richard Bauer (co-author: Ulrich Gebhardt), Clinic of Nuclear Medicine, Technical University München, Ismaninger Str. 22, DW-8000 München 80, Germany.

Richard Berberich, Department of Nuclear Medicine, University Clinic of Radiology, DW-6650 Homburg/Saar, Germany.

Hans J. Biersack, Institute of Nuclear Medicine, University of Bonn, Sigmund-Freud-Str. 25, DW-5300 Bonn 1, Germany.

Peter H. Cox, Department of Nuclear Medicine, Dr. Daniel den Hoed Clinic, P.O. Box 5201, 3008 AE Rotterdam, The Netherlands.

Maria Granowska (co-author: Keith Britton), Nuclear Medicine Department, St. Bartholomew's and Medical College, West Smithfield, London EC1A 7BE, U.K.

Klaus F. Gratz (co-authors: Otmar Schober and Burckhard Ringe), Department of Nuclear Medicine and Biophysics, Medical University Hannover, Konstanty-Gutschow-Str 8, DW-3000 Hannover 61, Germany.

Andreas L. Hotze (co-author: Hans J. Biersack), Institute of Nuclear Medicine, University of Bonn, Sigmund-Freud-Str. 25, DW-5300 Bonn 1, Germany.

F.D. Maul (co-authors: G. Hör, I. Brandhorst and R. Standke), Department of Radiology, Division of Nuclear Medicine, J.W. Goethe University, Theodor-Stern-Kai 7, DW-6000 Frankfurt am Main 70, Germany.

V. Ralph McCready, Consultant Nuclear Medicine and Ultrasound, The Royal Marsden Hospital, London & Surrey, Downs Road, Sutton, Surrey SM2 5PT, U.K.

W. Mecklenbeck (co-author: Henning Vosberg), Clinic for Nuclear Medicine, University of Düsseldorf, Moorenstr. 5, DW-4000 Düsseldorf 1, Germany.

Hee-Myung Park (co-authors: Henry N. Wellman and James A. Madura),

Division of Nuclear Medicine, University Hospital, P16, 926 W. Michigan Street, Indianapolis, IN 46202-5253, U.S.A.
Nicole A.M. Puttemans (co-authors: Pierre P. Andre, Serge A.M.Y. Jamsin, Daniel P.H. Balikdjian, and François Lustman), Department of Internal Medicine, Centre Hospitalier, Molière-Longchamp, Rue Marconi 142, B-1180 Brussels, Belgium.
Aslam R. Siddiqui (co-author: Henry N. Wellman), Division of Nuclear Medicine, University Hospital, P16, 926 W. Michigan Street, Indianapolis, IN 46202-5253, U.S.A.
Alan Siegel (co-author: Abass Alavi), Division of Nuclear Medicine, Hospital of the University of Pennsylvania, 3400 Spruce Street, Philadelphia, PA 19104, U.S.A.
Werner Waters, Institute for Nuclear Medicine, University of Cologne, Joseph-Stelzmann-Str. 9, DW-5000 Cologne 41, Germany.

PART ONE

Liver and Bile

1. Liver scintigraphy

V. RALPH McCREADY

The liver is the largest organ in the body, lying in the hypochondrium. Its importance lies in the many functions it carries out including carbohydrate metabolism, protein metabolism, lipid metabolism, excretion of haem following red cell breakdown, the production of bile salts to aid fat digestion, vitamin storage and metabolism. Weighing about 1200–1500 grams the bulk of the cells are hepatocytes with about 20% being reticuloendothelial cells. The reticuloendothelial cells (RES) of the liver make up about 80% of the total body's RES, line the sinusoidal tracts within the liver and are involved in the processing of red cells and phagocytosis of particulate matter in the blood such as immune complexes formed after antigen/antibody reactions. RES phagocytosis forms the basis of liver scintigraphy for the detection of space occupying lesions but liver imaging using radionuclides may also involve the display of other aspects of liver function such as the breakdown of red blood cells in various forms of anaemia, carbohydrate metabolism, biliary function, and the non specific metabolism of agents such as gallium and protein metabolism of amino acids such as methionine. The large blood supply of about 1500 ml per minute derived both from the systemic and portal circulations may explain the frequency of metastases from primary cancer located in the gut or elswhere. It also helps to explain the rapid uptake of many radiopharmaceuticals which are concentrated and excreted via the liver.

Technique

Liver imaging may be performed using a rectilinear scanner, the Anger gamma camera using either a planar or tomographic technique, or by using Positron Emission Tomography (PET). The choice of technique is determined by the information required and the radiopharmaceutical or metabolic process being studied. Although now outmoded for routine use the rectilinear scanner had the advantage of producing tomographic slices, the focussed collimators being insensitive to the activity in the superficial tissues. This

H.J. Biersack and P.H. Cox (eds), Nuclear Medicine in Gasteroenterology, 3–20

explains the unexpected relatively good results produced by that technique when compared with planar images produced by the Anger camera. In the case of the Anger camera the resolution falls off rapidly with depth resulting in only superficial lesions being seen with any clarity. The greatest problem in liver imaging where abnormalities are seen as cold areas is the lack of contrast between the lesion and the normal uptake of the radiopharmaceutical in the surrounding normal tissues. The low tumour to normal (T:N) contrast of approximately 2:1 found even when rectilinear scans are used means that only quite large abnormalities are visualised. Intuitively at first sight compared with ultrasound the radioisotope technique would seem to be of little value. However the main limiting factor in imaging tumours (or any other lesion) with ultrasound or any other modality is also the contrast between the abnormality and the surrounding tissue. This inevitably must vary from tumour to tumour producing a false negative rate which is very difficult to determine. It is also obvious that the factors which determine the signal from any particular lesion in each modality are different so that ultrasound or magnetic resonance imaging or X-ray CT scanning may each vary in their ability to visualise any individual abnormality.

Planar imaging

The quality of planar images is improved by placing the liver as close as possible to the collimator. Thus for best results a series of images around the right side of the patient from anterior to posterior should be taken. The problem of liver motion due to respiration can be minimised by imaging the patient in the erect position [1] but analogue motion correction is more successful [2]. The improvement in detection with motion correction has been found to be a function of the size of the lesion. It is important that an adequate number of photons be collected to reduce problems produced by counting statistics. It is usual to collect more than 500 K events in each view. Also since interpretation of liver images is difficult good quality control of the gamma camera is essential to ensure best uniformity, linearity and resolution.

Emission tomography

Emission tomographic imaging is a method of improving the contrast between normal and abnormal tissue and thus enables smaller lesions to be detected. In liver emission tomography the Anger gamma camera is rotated around the patient collecting a series of planar views. Usually 64 separate acquisitions are taken during a 360° rotation. Typically each acquisition takes 20 s and accumulates 70 K counts in that time after 80 MBq ^{99m}Tc colloid i.v. For best resolution it is important to keep the camera detector as close as possible to the patient. One way of doing this is to use an elliptical rotation. The advantage of elliptical rotation is that some artefacts associated with circular rotation tomography are eliminated. Special attention to quality control of the gamma camera is required as any non uniformity, non linearity or errors in reconstruction can result in artefacts in the final images which may be misinterpreted as pathology. For the liver a high resolution collimator is preferred but since this reduces sensitivity care has to be taken to collect enough photons to minimise statistical variation during reconstruction. Due to attenuation by the structures to the left of the liver it is preferable to record images through an arc of 180° centred around the liver [3]. This improves the contrast and enables smaller lesions to be identified. The problems in the interpretation of liver tomograms lies in the differentiation between non radioactive lesions and the mottle produced by the statistics of reconstruction. The ramp filter used in the reconstruction algorithm amplifies the noise at the high frequency end of the spectrum. It is important to use the correct filter therefore in liver studies. A ramp + linear 3D median window Hanning filter ($F = 0.83 \text{ cm} - {}^{1}$) appears to give a good compromise between sensitivity and mottle [4]. More sophisticated techniques can enhance the quality of tomographic images including constrained deconvolution [5].

PET imaging

Positron Emission Tomographic imaging (PET) is more usually used for brain and heart imaging. The potential advantages in the case of liver imaging are the improved resolution possible with this technique. This is due to the high resolution of the detectors using either BGO crystals or a multiwire proportional chamber (MWPC) and the use of the annihilation coincidence detection (ACD) of the photons which are emitted at 180° to each other when a positron meets an electron. The use of ACD to define the line where an event has occured gives uniform resolution through the body slice unlike single photon systems where the resolution decreases rapidly as the distance between the event and the collimator increases. The production of the positron labelled colloid can be achieved 'in house' using Gallium 68 from a long lived Germanium 68 generator [6]. With the proliferation of large MWPC devices larger area imaging will be possible enabling liver studies to be carried out both for localisation and metabolic studies.

Scintigraphic appearances

Normal liver scintigraphic appearances

The normal liver is quite variable in shape but most variations can be grouped into 12 categories [7]. The most common appearance is a triangular configur-

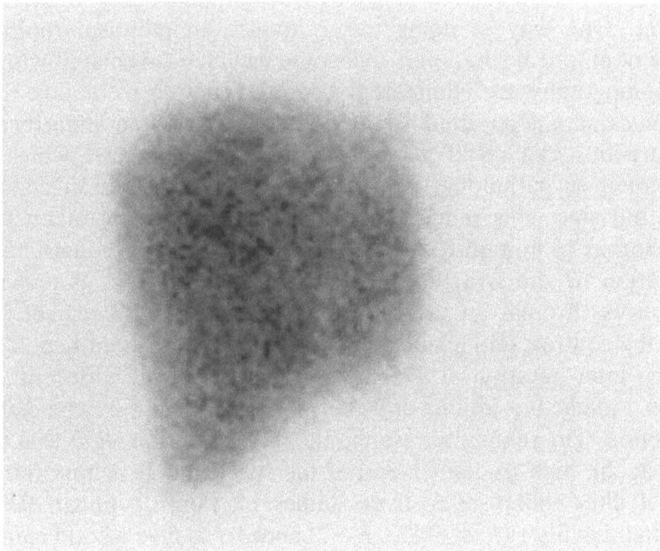

Figure 1. Anterior ^{99m}Tc colloid scintigram of a liver with an absent left lobe.

ation (65%). There is a prominent dome with an elevated diaphragm in about 13% of patients. The liver is divided into the smaller left lobe and the larger right lobe. The right and left lobes are separated by a fissure which may or may not be visible on a scintigram. The left lobe is sometimes thinned or may even be absent (Fig. 1) while the right lobe may be elongated below the costal margin with the so called 'Riedels Lobe' appearance seen in about 5% of patients. An example of this appearance is seen in Fig. 2. In about 15% of patients there is a prominent umbilical notch. The gall bladder may also produce a notch on the inferior edge of the liver border. The porta hepatis lies to the right of the fissure. Enlargement of the bile ducts or lymph glands at the porta hepatis may be seen as a photon deficient area. Occasionally the right lower costal margin can make an impression on the lateral aspect of the right lobe. The posterior view of the liver is also triangular in shape with often an area of reduced activity medially due to the renal impression. In the midline there is a vertical band of reduced activity due to the attenuation of the low energy photons of ^{99m}Tc by the bone mineral of the spine. It is unusual to see the left lobe of the liver in the posterior view. An awareness of these variations is important to avoid false positive diagnoses of space occupying lesions. In addition metallic objects or even soft tissue such as a large breast or roll of fat may produce a photon deficient area on scintigraphy. The spleen is seen most clearly in the posterior view due to its position close to the posterior surface of the abdomen.

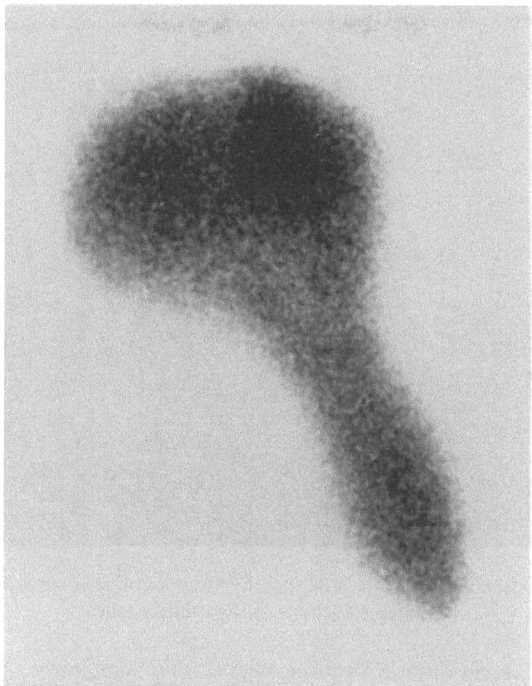

Figure 2. Lateral colloid scintigram of a liver with a prominent Reidels lobe.

Abnormal scintigraphic appearances

Abnormal appearances can be conveniently divided into lesions which have increased or decreased activity relative to the surrounding normal tissue.

Increased activity. Causes of increased activity include haemangioma of the liver, hepatoma, injection of radiocolloid in the hepatic vein [9, 10], superior or inferior vena cava obstruction [8], or both [11], and thrombosis of the intrahepatic veins [12, 13]. The obstruction of the venous outflow of the liver results in the Budd Chiari syndrome where the patient complains of abdominal pain, tender hepatomegaly and has signs of ascites. In the liver scintigram there is often increased activity in the position of the caudate lobe (Fig. 3), but occlusion of the various parts of the venous drainage system of the liver can cause other patterns of increased uptake [14].

Decreased activity. Most abnormalities of the liver produce photon deficient areas since the colloid concentrating RES cells are absent. These can be conveniently grouped as follows:

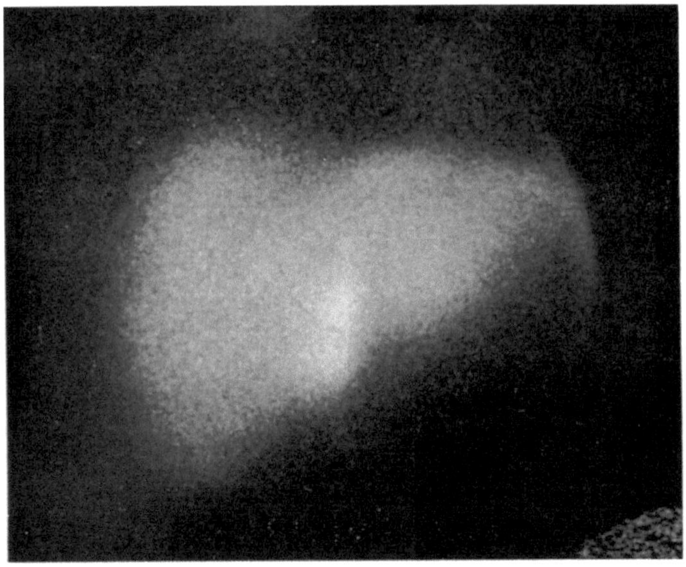

Figure 3. Anterior colloid scintigram in a patient with superior vena caval obstruction showing increased activity in the caudate lobe.

Causes of Photon Deficient Areas in Liver Scintigraphy	
Congenital	Cystic Disease
Infective	Abscesses Hydatid Cysts Granulomas
Traumatic	Haematoma Laceration Post Surgery
Neoplastic Benign	Adenoma Haemangioma Lipoma Focal Nodular Hyperplasia
Malignant	
Primary	Liver Cell Carcinoma Cholangiosarcoma
Secondary	Metastases Leukaemia Lymphoma

Differential diagnosis

There is little on the scintigrams which may lead the diagnostician to a differential diagnosis on liver scintigraphy alone. Figure 4a is the scintigram

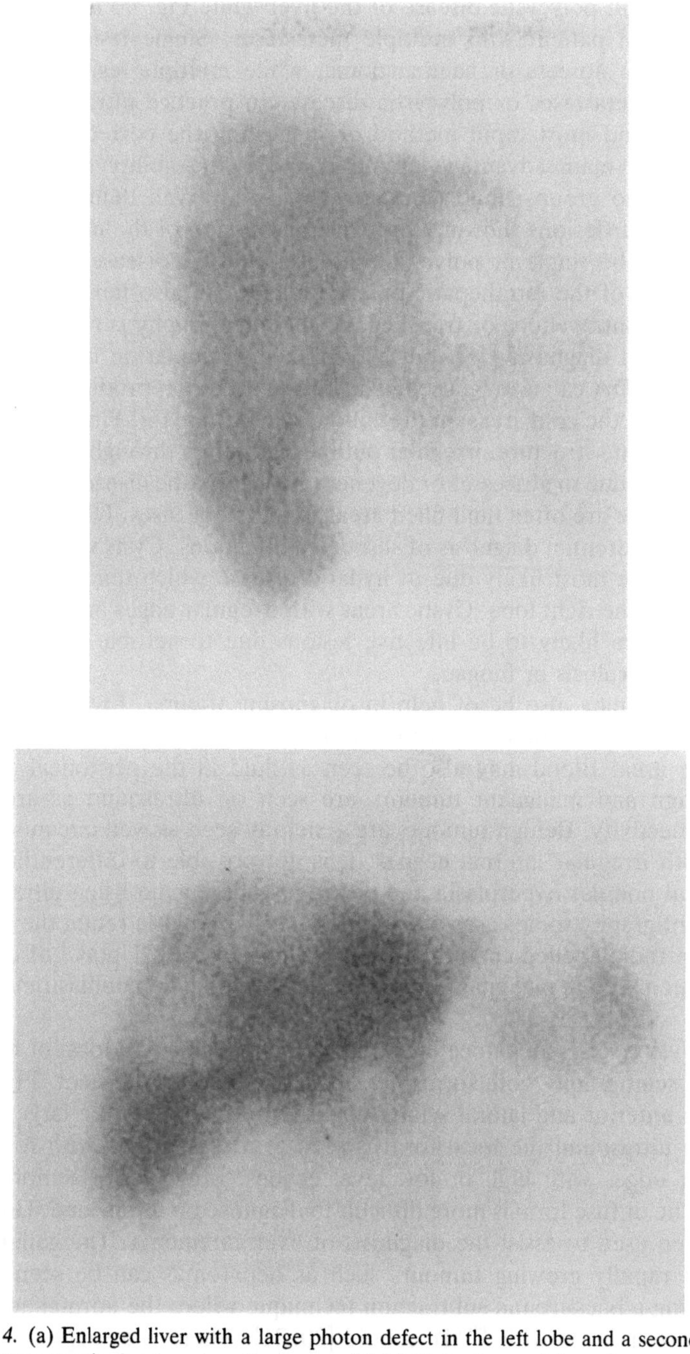

Figure 4. (a) Enlarged liver with a large photon defect in the left lobe and a second involving the lower margin due to polycystic disease. (b) Multiple photon defects due to metastases.

of a patient with polycystic disease of the liver while Fig. 4b is the anterior scintigram of a patient with multiple metastases. Single lesions favour a diagnosis of an abscess or haemangioma, while multiple lesions favour a diagnosis of metastases or polycystic disease. In practice ultrasound is the best, easiest and most rapid method of suggesting the correct differential diagnosis. The main advantage of ultrasound is the ability to divide the lesions into two groups: fluid containing, or solid. Well defined single or multiple circular lesions showing through transmission of the ultrasound are most likely to be single or polycystic disease. Caroli's disease where there are dilatations of the intrahepatic biliary radicles can also mimic polycystic disease. Here intravenous or transhepatic cholangiography is recommended to make a firm diagnosis [15], but a combined examination using labelled colloid and HIDA can also be used. The biliary excretion produces hot spots coinciding with the cold areas on the colloid scintigram [16]. Fluid containing lesions with some structure, irregular outlines and some through transmission are likely to be due to abscesses or degenerative neoplastic disease. In ovarian carcinoma there are often fluid filled areas in the metastases. Ultrasound can help in the differential diagnosis of suspected infections. Cysts with daughter cysts within are most likely due to hydatid disease which tends to be seen more often in the right lobe. Cystic areas with irregular edges and containing debris are more likely to be infective lesions due to aerobic or anaerobic bacteria, tuberculosis or fungae.

Ultrasound may also be of help in diagnosing trauma. Liver damage is seen as areas with low level echoes while blood may be seen as a fluid containing area. Blood may also be seen as fluid in the peritoneal cavity. Both benign and malignant tumours are seen on ultrasound as areas of altered reflectivity. Benign tumours are generally seen as well circumscribed masses with irregular internal echoes. It is not possible to differentiate between focal nodular hyperplasia and oestrogen adenomata using ultrasound but on scintigraphy some cases of focal nodular hyperplasia retain the ability to take up radiolabelled colloid [17] and gallium 67 [18]. Uptake of colloid has also been seen in malignant liver tumours including hepatoblastoma [19].

Primary liver cancer. Hepatocellular carcinoma presents in a focal or diffuse form. On scintigraphy both forms are imaged as photon deficient. Figure 5 shows the anterior and lateral scintigrams from a patient with a large hepatoma. On ultrasound the focal form is seen as a mass lesion with rounded or lobular edges with high or low level echoes, often with haemorrhagic regions. The diffuse form is more difficult to diagnose on ultrasound. Gallium 67 has been used to assist the diagnosis of liver carcinoma. The gallium 67 uptake in rapidly growing tumours such as hepatomas can be seen more clearly using a background subtraction technique where the normal areas of the liver are labelled with ^{99m}Tc colloid [20, 21]. There is a suggestion that the gallium uptake is related to the histology with increased uptake in moderate to well differentiated lesions unless there is reduced perfusion or there

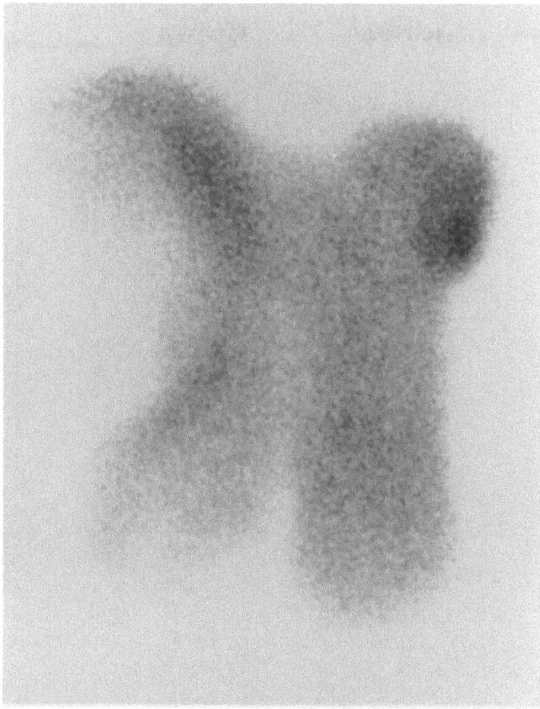

Figure 5a.

is significant necrosis [22]. Anecdotally other radiopharmaceuticals have been found to concentrate in hepatocellular carcinoma including ^{99m}Tc (Sn)-N-pyridoxyl-5 methyltryptophan [23], ^{99m}Tc Pyridoxylidene isoleucine [24], ^{99m}Tc Pyridoxylidene Glumate [25], and ^{75}Se selenomethionine [26]. However agents which reflect the various aspects of normal or neoplastic metabolism are not specific and uptake of methionine has also been seen in a hyperplastic nodule of the liver [27]. Failure of selenomethionine to concentrate in primary liver cancer in patients from South Africa has been attributed to the anaplastic nature and frequency of necrosis in the type of tumour which is found in the coloured population [26]. The increased metabolism seen in neoplasms compared with normal liver has been demonstrated by PET imaging of ^{13}N ammonia in hepatocellular carcinoma [28]. Some retention of hepatocyte function has been reported by Lee and Shapiro [29], who found ^{99m}Tc HIDA function in both the primary tumour and distant metastases.

Metastatic disease. Liver metastatic disease is probably the most frequent diagnostic problem encountered in liver scintigraphy. Most primary tumours produce secondary lesions in the liver at some time during the course of the disease. The scintigraphic diagnosis of early liver disease remains difficult in

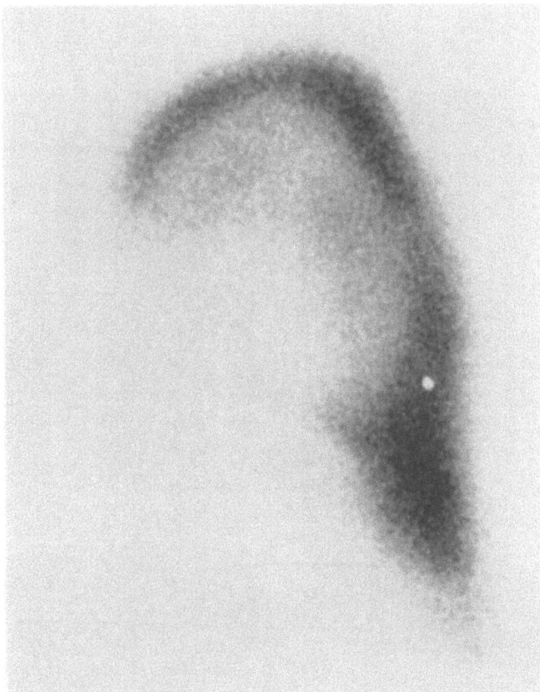

Figure 5. a, b: Anterior and lateral views of a colloid liver scintigram with a large filling defect due to a primary hepatoma.

spite of technical advances over the past few years. The difficulty in diagnosis is not confined to radioisotope methods. Studies in patients with breast carcinoma comparing the last liver in vivo or in vitro diagnostic test, with subsequent post mortem results have shown a wide disparity. Using colloid scintigraphy it is thought that the minimum size of lesion detectable is about 2 cm superficially, while deep lesions need to be at least 3 cm before they can be seen. By including features such as hepatomegaly, irregular distribution of tracer, as well as the detection of cold areas, the sensitivity, specificity and accuracy can be as high as 90% but of course the false positive rate rises [30]. Emission tomography helps to improve the detection of lesions with a range of sensitivities and specificities between 80–94% being quoted [31, 32]. To achieve good sensitivity it is necessary to be aware of and prevent the many causes of false positive diagnoses. These include metallic objects overlying the liver such as braces, belts, buttons etc. The low energy photons from ^{99m}Tc are absorbed by metal or other objects. Likewise a pendulous breast can simulate a space occupying lesion in the upper part of the right lobe. The variations in outline have already been mentioned. Lesions at the edge of the liver can be easily misinterpreted as normal variation. Photon deficient

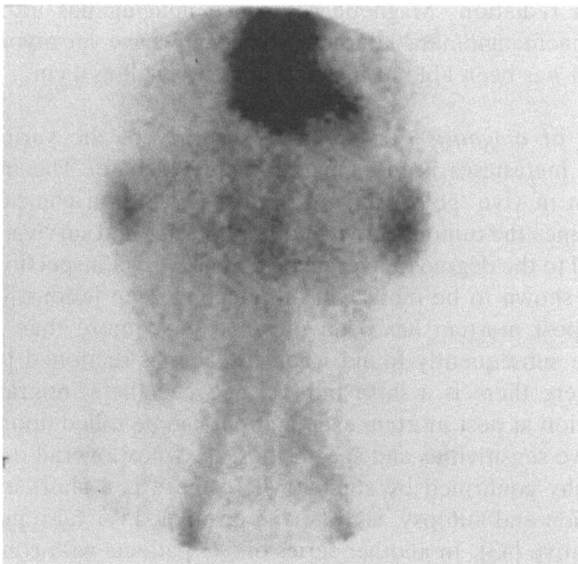

Figure 6. An early image from a dynamic study following an intravenous injection of ^{99m}Tc labelled autologous red blood cells. The increased activity in the upper part of the right lobe laterally is due to a haemangioma. The vacular space in the spleen and heart are clearly visible.

areas in the region of the gall bladder and porta hepatis can be evaluated using ^{99m}Tc HIDA [33, 35].

In the further investigation of liver metastatic disease ultrasound is again the most convenient further investigation. Metastases may be seen as areas with either reduced or increased echo patterns. They may be solid or contain fluid filled areas due to degeneration. The fluid containing areas have to be differentiated from benign cystic disease. Echogenic lesions include benign haemangioma of the liver. Although they are often echo poor, thrombosis or calcification produces high level echoes. This calcification does not normally concentrate bone imaging agents with exceptions reported by Burkhalter [35].

Haemangiomata are commoner in the age group where breast carcinoma is most frequent, so a single echogenic lesion can pose real diagnostic difficulties [36]. Scintigraphy using in vivo ^{99m}Tc labelled red cells is quick and effective for confirming haemangiomata [37]. Figure 6 shows an image during the intravenous injection of autologous ^{99m}Tc labelled red blood cells. The haemangioma can be seen in the upper part of the right lobe laterally. The delayed study shows haemangiomas as an area with increased local blood volume [38].

Emission computed tomography improves the visibility of such lesions. CT scanning using contrast, or angiography is also used to give a definitive diagnosis [39], but of course the radioisotope method is less invasive and gives less radiation. Magnetic resonance imaging has also been used to confirm haemangiomata although some may have an atypical appearance [40]. MRI has been able to image lesions as small as 1 cm.

Accuracy of diagnosis. The relative accuracy of the various methods of detecting metastases is the subject of great debate. The main problem is finding an in vivo 'gold standard.' Series relying on clinical follow up are suspect, since the tumours grow at different rates and survival is not necessarily related to the degree of liver involvement. Direct inspection at laparotomy has been shown to be inefficient at detecting deep lesions [41], while palpation at post mortem has been shown to miss more than 10% of lesions which are subsequently found when the liver is sectioned [42]. Thus, only series where there is a short interval between the scintigram or scan, and confirmation at post mortem examination, can be relied upon when comparing relative sensitivities and specificities. A typical overall detection rate for scintigraphy confirmed by studies where there is a short interval between examination and autopsy, is 81% true positive, 15% false positive and 21% false negative [43]. In another series of 581 patients with confirmation made within 40 days, the overall accuracy was 77.3% [44]. The accuracy of detection depends upon the type of primary being studied. Gut lesions tend to be single and well defined and therefore more easily visualised. In one series of patients with colonic carcinoma 88% of lesions were correctly detected [45]. In breast carcinoma, where the metastases tend to be smaller and more diffuse, the figure was 67%. It is generally felt that ultrasound is more sensitive than radioisotope scintigraphy while X-ray computed tomography has the highest sensitivity and specificity (91% and 96% respectively in a series reported by Brendel [46]). An exception is in the detection of carcinoid liver metastases, where colloid scans are felt to be superior to X-ray CT scanning [47].

In summary for the detection of space occupying lesions liver scintigraphy gives the overall view of the liver, but deep lesions are difficult to image. Ultrasound has high resolution throughout the slice and can usually detect small lesions at any depth, but in many patients parts of the liver are obscured by gas in the gut, or by overlying ribs. X-ray CT does not have either of these problems but is somewhat more time consuming and expensive. However it is the most sensitive technique.

The place of newer techniques for the differential diagnosis of solid lesions has yet to be assessed. CEA antibodies have been used in digestive tract cancer involving the liver with a 70% correct detection rate with only 1% false positives [48]. The low blood perfusion often associated with tumours will probably limit attempts to improve antibody imaging to the level of a routine diagnostic test. There have been several reports of diphosphonate [49] or similar compounds, being taken up in secondary cancer of the liver [50]. This does not seem to be related to calcification but may represent pyrophosphate binding by collagen [51]. However, Fluorine 18 has also been

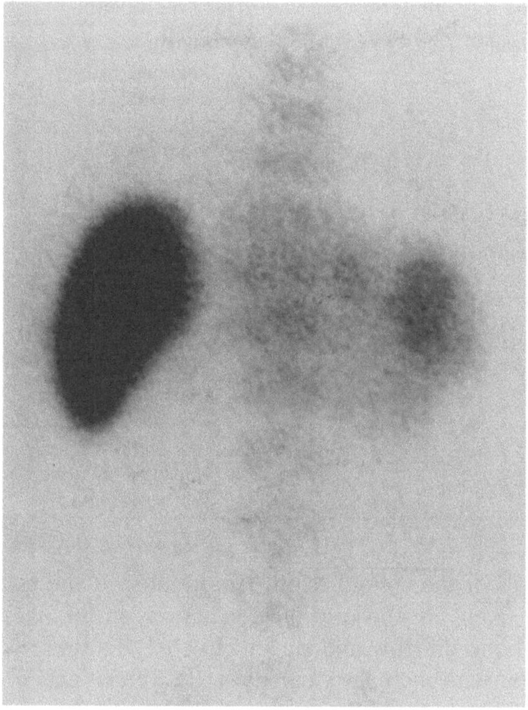

Figure 7. A posterior view of a colloid scintigram in a patient with advanced cirrhosis. The uptake in the liver is reduced with increased activity in the bone marrow and the spleen.

reported to concentrate in colon metastases. Of 15 patients, seven showed localisation of the Fluorine [52].

Diffuse disease. Most diffuse diseases produce a generalised alteration in the uptake of the colloid in the liver. Since the colloid is distributed between the bone marrow, liver, and spleen, a reduction in liver function results in increased uptake in the bone marrow and spleen. The degree of redistribution is related to the level of liver damage and reflects clinical improvement or deterioration. Figure 7 shows the posterior scintigram of a patient with advanced cirrhosis of the liver. The uptake in the bone marrow can be seen while the activity in the liver is decreased. Techniques for demonstrating diffuse liver disease are similar to those for focal disease, but if quantitation is required, then emission computed tomography is valuable since the volume and fractional uptake of the colloid in the several organs can be measured [53]. However, planar imaging with some form of computer analysis is more commonly used [54].

Causes of diffuse abnormalities of the liver include		
Infective		Malaria Mononucleosis Schistosomiasis Syphilis Weils disease Hepatitis
Degenerative	Cirrhosis	
Neoplastic	Benign Malignant	– Leukaemia Lymphoma
Metabolic		Amyloidosis Fatty infiltration Gauchers disease Neimann–Pick disease von Gierke's disease Haemachromatosis Wilsons disease

The appearances of the colloid scintigram in most of the established diffuse diseases of the liver, are those of uneven activity in the liver with increased activity in the bone marrow and spleen. In cirrhosis there is initial enlargement of the liver with increased activity in the spleen and bone marrow. As the disease progresses there is atrophy of the right lobe, and compensatory hypertrophy of the left lobe. Any disease producing long term fibrosis will produce similar appearances which are related to the extraction efficiency for colloid, and the blood flow to the various parts of the liver. When cirrhosis is induced in rats there is a reduction in the extraction efficiency for colloid associated with early histological changes. When regeneration occurs the extraction efficiency returns. In the long term both the blood flow and extraction efficiency drops [55].

In patients with alcoholic liver disease the displacement of uptake to the extrahepatic regions is impressive. In approximately 40% of 90 patients with alcoholic liver disease, the vertebrae are as intensely imaged as the liver, while 35% have similar levels in the ribs (56). Measurements of the right to left lobe activity ratio, using quantitative techniques, have been found to be more sensitive and specific for alcohol cirrhosis than other criteria [57]. Rectal Thallium 201 has also been used to measure the severity of cirrhosis. The 25th minute ratio of heart to liver activity is used as an index of portal shunting. The ratio is normal in alcoholic patients, slightly higher with fibrosis and significantly higher with cirrhosis [58].

Metabolic diseases such as von Gierke's disease, amyloidosis, result in fatty infiltration of the liver. On colloid scintigraphy this is seen as hepatomegaly with or without uneven distribution of the radioactivity. Ultrasound is of little help in the differential diagnosis of these conditions. The main finding in fatty liver is an increase in the liver parenchymal echoes. The liver acquires a uniform appearance with an apparent loss of internal structure. However, the liver may be normal although the disease is severe. In cirrhosis extra features may be apparent, such as irregularity of the surface of the liver, hypertrophy of the caudate lobe, ascites and signs of portal hypertension including recanalisation of the portal vein, ascites and dilatation of the splenic vein [59].

Conclusion

Although the sensitivity of liver colloid scintigraphy is surprisingly high, and the technique relatively simple, in most cases ultrasound remains the first best technique for the detection and differential diagnosis of space occupying lesions of the liver. However, scintigraphy is useful for an overall assessment of liver size and for RES function and distribution. Newer techniques using radio pharmaceuticals such as labelled methionine fluorodeoxyglucose as markers of cellular division and metabolism, show promise for the evaluation of tumour growth and response to therapy. The improved resolution available possible with PET should ensure a continued role for radioisotope studies of liver morphology and function.

References

1. Mettler Jr FA, Shea Jr WH, Guiberteau MJ, Potsaid MS (1977) 'Improvement in visualization of hepatic lesions with upright view.' *J Nucl Med* 18: 1128–1130.
2. Haraux G, Bronskill MJ (1979) 'Comparison of the liver's respiratory motion in the supine and upright positions: concise communication.' *J Nucl Med* 20: 733–735.
3. Ott RJ, Flower MA, Khan O, Kalirai T, Webb S, Leach MO, McCready VR (1983) A comparison between 180 and 360 data reconstruction in single photon computed tomography of the liver and spleen.' *Br J Radiol* 56: 931–937.
4. Webb S (1985) 'Comparison of data processing techniques for the improvement of contrast in SPECT liver tomograms.' *Physics in Medicine and Biology* 30: 1077–1086.
5. Webb S, Long A, Ott RJ, Leach MO, Flower MA (1985) 'Constrained deconvolution of SPECT liver tomograms by direct digital image restoration.' *Medical Physics* 12: 53–58.
6. Kumar B, Miller TR, Siegel BA, Mathias CJ, Markham J, Ehrhardt GJ, Welch MJ (1981) 'Positron tomographic imaging of the liver: ^{68}Ga iron hydroxide colloid.' *Am J Roent* 136: 685–690.
7. McAfee JG, Ause RG, Wagner HN Jnr (1965) 'Diagnostic value of scintillation scanning of the liver.' *Arch Int Med* 116: 95.
8. Gooneratne NS, Buse MG, Quinn JL, Selby JB (1977) ''Hot spot' on hepatic scintigraphy and radionuclide venacavography.' *Am J Roent* 129: 447–450.
9. Chhabria PB, Chandnani PC (1977) ''Hot spot' on radiocolloid scan of the liver.' *Clin Nucl Med* 1: 258–259.
10. Tetalman MR, Kusumi R, Gaughran G, Baba N (1978) 'Radionuclide liver spots: indicator of liver disease or a blood flow phenomenon.' *Am J Roent* 130: 291–296.
11. Desai AG, Park CH (1983) 'Cavo-portal shunting in superior and inferior vena caval obstruction.' *Clin Nucl Med* 8: 365–368.

12. Dhawan VM, Sziklas JJ, Spencer RP (1978) 'Pseudo-Budd–Chiari syndrome.' *Clin Nucl Med* 3: 30–31.
13. Hanelin LG, Uszier JM, Sommer DG (1975) 'Liver scan 'hot spot' in hepatic veno-occlusive disease.' *Radiology* 117: 637–638.
14. Picard M, Carrier L, Chartrand R, Franchebois P, Picard D, Guimond J (1987) 'Budd–Chiari syndrome: typical and atypical scintigraphic aspects.' *J. Nucl. Med.* 28: 803–809.
15. Sty, JR, Sullivan P, Wagner R, Starshak RJ (1978) 'Hepatic scintigraphy in Caroli's disease.' *Radiology* 127: 732.
16. Georgiou E, Alevizaki C, Proukakis C. (1983). 'Preoperative scintigraphic evaluation of the liver and biliary tract in Caroli's disease.' *Eur J Nucl Med* 8: 34–36.
17. Casarella W, Knowles D, Wolff M, Johnson P (1978) 'Focal nodular hyperplasia and liver cell adenoma.' *Am J Radiol* 131: 393–402.
18. Belanger MA, Beauchamp JM, Neitzschman HR (1975) 'Gallium uptake in benign tumor of liver: case report.' *J Nucl Med* 16: 470–471.
19. Diament MJ, Parvey LS, Tonkin ILD, Johnson KD, Bernstein R, Webber, B (1982) 'Hepatoblastoma: technetium sulfur colloid uptake simulating focal nodular hyperplasia.' *Am J Roent* 139: 168–171.
20. Buraggi GL, Laurini R, Rodari A, Bombardieri E (1976) 'Double-tracer scintigraphy with ^{67}Ga citrate and ^{99m}Tc sulfur colloid in the diagnosis of hepatic tumors.' *J Nucl Med* 17: 369–373.
21. Muller-Brand J, Benz U, Kyle CA Boss M, Fridrich R (1977) 'Triple radioisotope technique in etiologic evaluation of space-occupying lesions of the liver.' *Eur J Nucl Med* 2: 231–238.
22. Waxmann AD, Richmond R, Juttner H, Siemsen JK, Heffelinger MJ, Fink E (1980) 'Correlation of contrast angiography and histologic pattern with gallium uptake in primary liver-cell carcinoma: non-correlation with alpha-feto protein. Concise communication.' *J Nucl Med* 21: 324–327.
23. Hasagawa Y, Nakano S, Ibuka K, Hashizume T, Sasaki Y, Imaoka S, Ishiguro S, Kasugai H, Okano Y, Tanaka, S, Ehara M, Morii T, Kojima J, Isigami S. (1984) 'The importance of delayed imaging in the study of hepatoma with a new hepatobiliary agent. *J Nucl Med* 25: 1122–1126.
24. Ueno K, Haseda Y (1980) 'Concentration and clearance of ^{99m}Tc-pyridoxylidene isoleucine by a hepatoma.' *Clin Nucl Med* 5: 196–199.
25. Utz JA, Lull RJ, Anderson JH, Lambrecht RW, Brown JM, Henry W (1980) 'Hepatoma visualization with ^{99m}Tc pyridoxylidene glutamate.' *J Nucl Med* 21: 747–749.
26. Kew MC, Geddes EW, Levin J (1974) 'False-negative ^{75}Se-selenomethionine scans in primary liver cancer.' *J Nucl Med* 15: 234–236.
27. Douglas JG, Zambartas CN, Sumerline MD, Finlayson NDC (1981) '^{75}Selenomethionine in the diagnosis of hepatocellular carcinoma. Report of a false positive scan.' *Eur J Nucl Med* 6: 91–92.
28. Hayashi N, Tamaki N, Yonekura Y, Senda M, Saji H, Yamamoto K, Konishi J, Torizuka K (1985) 'Imaging of the hepatocellular carcinoma using dynamic positron emission tomography with nitrogen-13 ammonia.' *J Nucl Med* **26**: 254–257.
29. Lee VW, Shapiro JH (1983) 'Specific diagnosis of hepatoma using ^{99m}Tc-HIDA and other radionuclides.' *Eur J Nucl Med* 8: 191–195.
30. Galli G, Maini CL, Salvatori M, Ausili Cefaro G (1982) 'The diagnostic application of radiocolloid liver scintigraphy in breast carcinoma.' *Nucl Med* 21: 140–144.
31. Van Heertum RL, Brunetti JC, Yudd AP (1987) 'Abdominal SPECT imaging.' Seminars in *Nucl Med* 17: 230–246.
32. Berche C, Aubry F, Langlais C, Vitaux J, Parmentier Cl, Di Paola R (1981) 'Diagnostic value of transverse axial tomoscintigraphy for the detection of hepatic metastases: results on 53 examinations and comparison with other diagnostic techniques.' *Eur J Nucl Med* **6**: 435–452.
33. Rao BK, Pastakia B, Lieberman LM (1980) 'Evaluation of focal defects on ^{99m}Tc sulfur colloid scans with new hepatobiliary agents.' *Radiology* 136: 497–499.

34. Schulze PJ, Stritzke P, Stolzenbach G (1981) 'Liver imaging and detection of liver metastases with ^{99m}Tc-HIDA.' *Nucl Med* 20: 214–219.
35. Burkhalter JL, Morano JU, Patel BR (1986) 'Accumulation of ^{99m}Tc MDP in a cavernous hemangioma of the liver.' *Clin Nucl Med* 11: 498–500.
36. Wiener SN, Parulekar SG (1979) 'Scintigraphy and ultrasonography of hepatic hemangioma.' *Radiology* 132: 149–153.
37. Engel MA, Marks DS, Sandler MA Shetty P (1983) 'Didderentiation of focal intrahepatic lesions with ^{99m}Tc-red blood cell imaging.' Radiology 146: 777–782.
38. Front D, Royal HD, Israel O, Parker JA, Kolodny GM (1981) 'Scintigraphy of hepatic hemangiomas: the value of ^{99m}Tc-labeled red blood cells: concise communication.' *J Nucl Med* 22: 684–687.
39. Freeny PC, Vimont TR, Barnett DC (1979) 'Cavernous haemangioma of the liver: ultrasonography, arteriography, and computed tomography. *Radiology* 132: 143–150.
40. Brown RKJ, Gomes A, King W, Pusey E, Lois J, Goldstein L., Busettil RW, Hawkins RA (1987) 'Hepatic hemangiomas: evaluation by magnetic resonance imaging and ^{99m}Tc red blood cell scintigraphy.' *J Nucl Med* 28: 1683–1687.
41. Golicher JC (1941) 'The operability of carcinoma of the rectum.' *Br Med J* ii–393.
42. Ozardo A, Pickren J. (1962) 'The topographic distribution of liver metastases: its relation to surgical and isotope diagnosis.' *J. Nucl. Med.* 3: 149.
43. Ostfeld DA, Meyer JE (1981) 'Liver scanning in cancer patients with short-interval autopsy correlation.' *Radiology* 138: 671–673.
44. Lunia S, Pathasarathy KL, Bakshi S, Merrill A (1975) 'An evaluation of ^{99m}Tc-sulfur colloid liver scintiscans and their usefulness in metastastic workup: a review of 1,424 studies.' *J Nucl Med* 16: 62–65.
45. Drum DE, Beard JM (1976) Scintigraphic criteria for hepatic metastases from cancer of the colon and breast.' *J Nucl Med* 17: 677–680.
46. Brendel AJ, Leccia F, Drouillard J, San Galli F, Eresue J, Wynchank S, Barat, J-L, Ducassou D (1984) 'Single photon emission computed tomography (SPECT), planar scintigraphy, and transmission computed tomography: a comparison of accuracy in diagnosing focal hepatic disease.' *Radiology* 153: 527–532.
47. Halvorsen RA, Wilkinson RH, Feldman JM (1987) 'Carcinoid liver metastases. Accuracy of radionuclide liver/spleen imaging compared to computed tomography. *Clin Nucl Med* 12: 268–273.
48. Aburano, T., Tonami, N., Hisada K (1979) 'Radioimmunoassay for carcinoembryonic antigen as an adjunct to liver scan in the detection of liver metastases from digestive-tract cancer.' *J Nucl Med* 20: 232–235.
49. Ghaed N, Marsden RJ (1978) 'Accumulation of ^{99m}Tc-diphosphonate in hepatic neoplasm.' *Radiology* 126: 192.
50. Shih W-J, Domstad PA, Lieber A, DeLand FH, Coupal JJ (1986) 'Localization of ^{99m}Tc HMDP in hepatic metastases from colonic carcinoma.' *Amer J Roent* 146: 333–336.
51. Stevens JS, Clark EE (1977) 'Liver metastasis of colon adenocarcinoma demonstrated on ^{99m}Tc-pyrophosphate bone scan.' *Clin Nucl Med* 2: 270–271.
52. Garcia AC, Yeh SDJ, Benua RS (1977) 'Accumulation of bone-seeking radio-nuclides in liver metastasis from colon carcinoma.' *Clin Nucl Med* 2: 265–269.
53. Kodama T, Watanabe K, Hoshi H, Jinnouchi S, Arakawa K, Kusumoto S, Honda H, (1988) 'Diagnosis of diffuse hepatocellular diseases using SPECT.' *J Nucl Med* 27: 616–619.
54. Wasnich R, Glober G, Hayashi T, Vicher T, Yeh F (1979). 'Simple computer quantitation of spleen-to-liver ratios in the diagnosis of hepatocellular disease. *J Nucl Med* 20: 149–154.
55. Goeting NLM, Fleming JS, Gallagher P, Walmsely BH, Karran SJ (1986) 'Alterations in liver blood flow and reticuloendothelial function in progressive cirrhosis in the rat.' *J Nucl Med* 27: 1751–1754.
56. Prakash V, Lin MS, Kriss JP (1977) 'Liver scintigraphy in alcoholic liver disease.' *Clin Nucl Med* 2: 308–309.
57. Shreiner DP, Barlai-Kovach M (1981) 'Diagnosis of alcoholic cirrhosis with the right-to-left hepatic lobe ratio: concise communication. *J Nucl Med* 22: 116–120.

58. Urbain D, Reding P, Georges B. *et al.* (1986) 'The clinical value of ^{201}Tl per rectum scintigraphy in the work-up of patients with alcoholic liver disease.' *Eur J Nucl Med* 12: 267–270.
59. Cosgrove DO (1988) 'The application of ultrasound in surgery of the liver.' In Blumgart (ed), *Surgery of the Liver and Biliary Tract*, p. 167. Churchil 1 Livingstone.

2. Differential diagnosis of jaundice with hepatobiliary scintigraphy

F. D. MAUL, G. HÖR, I. BRANDHORST and R. STANDKE

Introduction

Hepatobiliary scintigraphy (HBS) is a well established non invasive procedure for the differential diagnosis of intra- and extrahepatic jaundice. Clinical impact of hepatobiliary scintigraphy increased since the introduction of the ^{99m}Tc labeled hepatobiliary imaging agents by Loberg and coworkers in 1975 [16, 31].

^{99m}Tc labeled HIDA offered the possibility to image hepatobiliary function and various pathophysiological conditions derived from the biokinetics concerning hepatic parenchyma and the bile duct system. Accordingly, gastroenterology, abdominal surgery, and paediatry can benefit from HBS, in diagnosis and follow up of jaundice.

This chapter is supposed to include reports from literature and is based on our own experience dating back more than 15 years [18].

Radiopharmacy and principal scintigraphic results

Development

The first in vivo test of liver function was introduced by Delprat in 1923 [11]. Delprat proposed bengal rose as a non radioactive labeled pharmaceutical. This type of in-vivo-test could only measure the blood disappearance of bengal rose after a fixed time. ^{131}I bengal rosa, as the first radionuclide labeled substance was introduced in 1955 [51]. Particularly because of its physical restrictions ^{131}I bengal rose was not suitable for clinical routine use and therefore did not achieve widespread acceptance.

A number of ^{99m}Tc labeled substances were proposed to overcome the problems inherent in the use of ^{131}I hepatobiliary radiopharmaceuticals [18, 27, 56]. Today the iminodiacetic acids (IDA) substances derived from lidocain became the radiopharmaceuticals of choice. ^{99m}Tc labeled hepatobiliary IDA (HIDA) substances were introduced by the group of Loberg in 1975

H.J. Biersack and P.H. Cox (eds), Nuclear Medicine in Gasteroenterology, 21–36

Table 1. Biokinetic pathways of IDA-derivatives.

Biokinetic function	From	To physiologic	To pathologic
Plasma clearance	Plasma	Hepatocytes	Kidneys
Biliary excretion	Hepatocytes	Bile ducts	Reentry → Plasma
Bile elimination	Bile ducts	Duodenum	Bile Leakage

[16, 31]. The ^{99m}Tc HIDA molecule has two different functional components [32]. One function uses a chelate group for the bond of ^{99m}Tc, the second molecular function is responsible for hepatic uptake, transport, and excretion [29, 39]. Even if uptake and transport mechanisms are not understood in detail it is generally accepted that HIDA derivatives are not metabolised during hepatic transit. In addition HIDA substances are not reabsorbed by the small intestine.

Biokinetics of HIDA substances

The biokinetics of HIDA substances can be described by 3 essential phases: (1) HIDA blood clearance depending mainly on the hepatic HIDA uptake, (2) hepatocellular excretion, and (3) duodenal bile elimination (Table 1). Besides these parameters different appearance times for instance in the bile ducts, gall bladder and duodenum are characterizing hepatobiliary functions in clinical routine. Different parameters for the same or similar split functions are used in literature to characterize HIDA biokinetics [20, 27, 42, 62]. Equivalent biokinetic parameters are summarized in Table 2.

Table 2. Equivalent biokinetic parameter of HIDA representing the 3 main split functions uptake, excretion and bile elimination.

Uptake
Plasma clearance constant ($k_{1/2}$)
Time of curve maximum (T_{max})
Extraction fraction[1]
Half-life time of uptake increase of the liver (HLU)[2]
Excretion
Appearance time in hepatic ducts
Excretion half-life time[1]
Mean parenchymal transit time[2]
Bile elimination
Appearance time in the duodenum
Percentage of activity retention after 40 min (RET_{40})[2]

[1]Krishnamurthy et al. 1990.
[2]Maul et al. 1982.

Different HIDA substances
Different HIDA radiopharmaceuticals were introduced. Widely used are diethy-IDA [46, 60], DISIDA [22], and paraisopropyl-IDA [38, 61]. DISIDA and diethyl-IDA show a high liver uptake and a fast transit. In the case of increased total bilirubin a part of the radiopharmaceutical is cleared by kidneys. Only parabutyl-IDA shares a neglectable amount of renal clearance but unfortunately it has a too long hepatic transit and excretion time [52, 10].

Halogenated HIDA. In the last decade new substances were developed for instance those with a halogen substitution. Substances of this new group of radiopharmaceuticals are trimethybrom IDA [24] and IODIDA [49, 50]. These new substances exhibit a comparable hepatic clearance and uptake as the above mentioned HIDA substances, however their renal clearance is very low.

HIDA blood clearance and hepatic uptake
Blood clearance of HIDA derivatives essentially is determined by their hepatic uptake which on his part is determined by liver perfusion as well as by hepatocellular HIDA extraction. In cirrhosis reduced liver uptake mainly depends on perfusion. This view is advocated by our experience that the reduction of hepatic blood flow due to the blockade of the portal vein under conditions of a preoperative shunt simulation results in significant impairment of hepatic uptake. Usually used uptake parameters are extraction fraction and T_{max}. A disadvantage of T_{max} is that it is a functional 'mixed' parameter depending on uptake and on clearance of parenchyma as well. Of course an impaired uptake function of the liver influences all subsequent functions, excretion and bile flow, which has to be considered in the evaluation of these parameters.

Uptake inhibition by bilirubin
An increase of total bilirubin inhibits the uptake of HIDA by competition [9, 17, 44]. As a consequence HIDA uptake may also severely inhibited in the case of jaundice. The uptake mechanism is comparable to that of bengal rosa.

Hepatocellular excretion of HIDA
Excretion is determined by 2 factors: hepatocellular transport function and the pressure in bile ducts. Different diseases can reduce the hepatocellular function of transport and excretion to varying degrees. In hepatocellular diseases the hepatic uptake function is also reduced.

Reduced excretion induced by an increased pressure in bile ducts
A high pressure in bile ducts reduces or stops the excretion. Immediately after the onset of obstruction the hepatic extraction function is not impaired

Table 3. Reference (normal) values of Frankfurt's parameter based on 52 individuals with a proven normal liver function.

Parameter	Mean normal	Normal Range
Half-life time of liver uptake in min (HLU)	2.5	1.7–3.0
Mean parenchymal transit time in min (MPTT)	39.0	21.0–51.0
Biliary activity retention after 40 min in %	11.5	6.8–14.9

[5, 21]. If an obstruction continues for a longer time the extraction function will be reduced although parenchyma is not the primary location of the disorder. If the time of jaundice due to an obstruction is short enough an imagable HBS with HIDA is possible. Typical signs of an incomplete obstruction are a reduction of excretion and a bile retention. Interestingly in all hepatic disorders the transit time or the excretion function is involved and for this reason reduction of excretion is a key parameter of all pathologic processes of the liver.

Excretion of halogenated HIDA in jaundice

Halogenated HIDA derivatives are newer imaging agents exhibiting no major renal elimination even if the total bilirubin levels are severely increased [49, 50]. Therefore this new class of HIDA derivatives is especially suitable for patients with severe jaundice.

Visualization of bile ducts in jaundice

Bile ducts normally can not be visualized in patients with a severe failure of excretion. However, these cases can easily be differentiated from a complete obstruction, often exhibiting no bile duct activity either. In the case of a severely impaired excretion we find no increase of biliary retention and an elimination of activity in duodenum can be seen in all cases. In cases with an incomplete obstruction, bile ducts can be mostly visualized [3].

Frankfurt's method

Previous multi-compartmental models proved to meet research purposes, but were unapplicable for clinical conditions. 3 parameters were introduced by our group to describe the 3 most important split functions of the hepatobiliary HIDA transport [35]. The basic idea for the development of these parameters was to have quick and easy to compute results with a high reproducibility and reliability avoiding complicated biokinetic models.

Reference (normal) values based on 52 subjects with a proven normal liver function are summarized in Table 3. A normal case is shown in Fig 1.

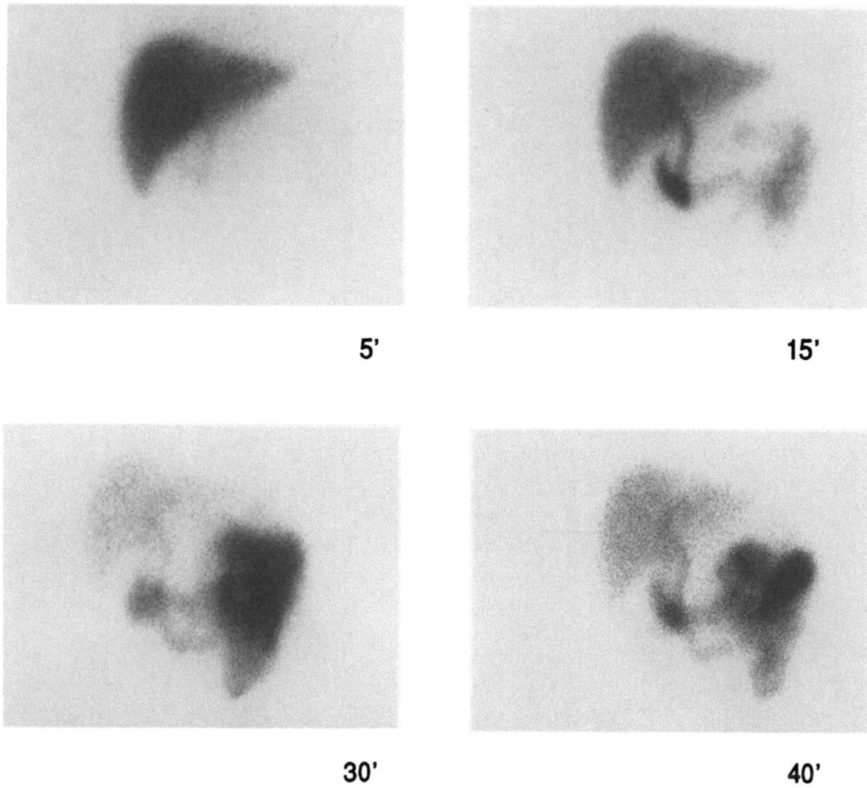

Figure 1. Case of normal hepatobiliary scintigraphy 5 min (upper left), 15 min (upper right), 30 min (lower right), and 40 min (lower left) after i.v. application of diethyl-HIDA. Nearly complete parenchymal clearance of the radiopharmaceutical after 40 min. Parameters were HLU: 1.5 min, MPTT: 23 min, RET_{40}: 5%.

Half-life time of liver uptake (HLU). As an equivalent for liver *uptake* we calculate the half-life time of liver uptake (HLU), i.e. the time in which the uptake increases by half. The program fits the total liver curve between 3 to 7 min after injection monoexponentially. Healthy persons have HLU values below 3 min. Different from the commonly used T_{max} this parameter is independent from the parenchymal clearance of HIDA.

Mean parenchymal transit time (MPTT). The *excretion* function is described by the *mean parenchymal transit time* (MPTT) characterizing the hepatobiliary passage of the radiopharmaceutical through the parenchyma. We have chosen this parameter because of the functional overlap of uptake and excretion being undistinguishable under pathological conditions within 40 min. Healthy persons present with mean parenchymal transit times below 50 min. This parameter correlates strongly with the half-life time of excretion –

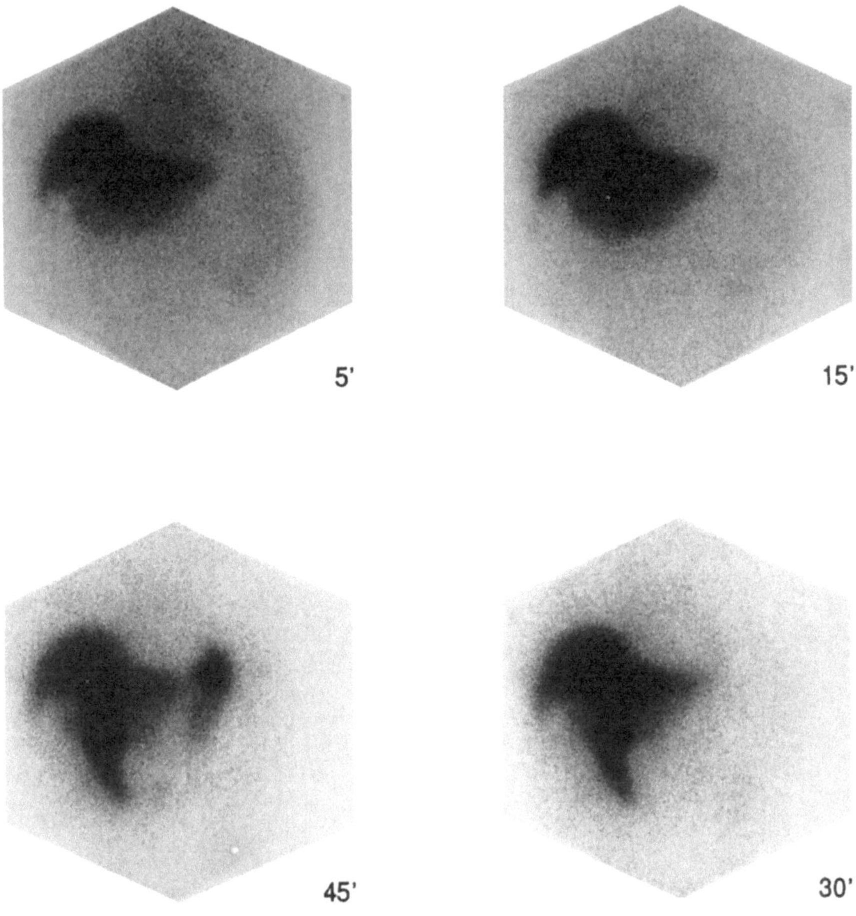

Figure 2. Hepatobiliary scintigraphy of a 50 years old male with parenchymal insufficiency due to an advanced cirrhosis. Times and arrangement of scintigrams are the same as in Fig. 1. HLU: 5.1 min, MPTT: 909 min, RET_{40}: 12%.

parameter which was proposed by Krishnamurthy and his group. Both parameters are derived from a monoexponential fit. We fit a time activity curve of a parenchymal ROI between 20 and 40 min.

Normalization of transit time. The effect of reduced uptake on excretion or transit times can be normalized [19, 33, 36]. In a study in which we investigated patients with cirrhoses before and after a diagnostic blockade of the portal perfusion – to simulate the postoperative situation of a portal shunt – we found that only those patients developed an anerobic metabolism who respond with a prolongation of normalized mean hepatic transit time whereas the group without anerobic reaction kept normalized mean hepatic transit times constant even if both groups indicated an increase of half life of uptake increase. Instead of normalization of the mean hepatic transit time we now prefer a multiparametric analysis.

Bile retention after 40 minutes. Finally *biliary activity retention* after 40 (RET_{40}) min is computed. This parameter is based on the Dost principle of corresponding areas which was developed for pharmacokinetic purposes. It is assumed that no essential amount of hepatic activity reenters plasma and that in the first 7 min no activity leaves the liver via bile ducts into duodenum. From the area under curve of a peripheral ROI which excludes visible hepatic bile ducts and covers the time from 0 to 7 min and from 7 to 40 min we calculate the activity which must be in the liver after 40 min if no activity has left the organ. This is a virtual 100% value of the total activity from which the part of retention in the hepatic bile ducts can be computed. For details see Maul et al. 1982 [35].

Multiparameter analysis. In the acute post hepatic blockade the hepatic uptake function – measured as half-life of uptake or as extraction fraction – is not disturbed because the increase of biliary pressure which stops excretion does not effect the uptake function directly. By our own experiences it is very helpful to consider the three split functions as half-life time of uptake, mean parenchymal transit time as a measure of excretion and the percentage of bile activity retention together which enables us to determine the underlying pathologic process. Similar procedures based on biokinetics for diagnostic decision making were proposed by other groups though with diagnostic algorithm based on quantitative [8, 23] as well as on mere qualitative evaluation of HBS [27].

Bifunctional blood clearance. Blood clearance curve represents a biexponential elimination mechanism. The initial phase of elimination parallels the early hepatic uptake whereas the delayed phase of the blood clearance correlates with the hepatic transit [33, 36].

Methods

The procedure of HBS can be divided in 3 different phases. *Phase 1* from injection up to 30 min. In this phase computer sequence should be acquired if a quantitative evaluation is planned. For qualitative evaluation scintigramms each 5 min are necessary. *Phase 2* lasts from 30 min up to 2 h (if necessary) and as long as there is remaining activity in bile ducts and liver. Scintigrams in this phase are acquired each 15 to 30 min. *Phase 3* Further scintigrams after 24 h are necessary in case activity persists or if no activity was detected in the duodenum.

Acquisition
For qualitative purposes scintigraphic standard methods are applied. For quantitative evaluation a computer acquisition is needed. Normally a 64 × 64 matrix will be used. The acquisition time per frame should be in the range of one minute. To evaluate the perfusion phase – for instance to separate the arterial and portal part of hepatic blood flow – shorter acquisition must be selected in the range of 1 s for the first minute.

Quantification
For quantitative analysis a study of 30 to 60 min is necessary depending on the program used. Our program needs 40 min which is tolerated by a jaundiced patient lying supine. The qualitative evaluation is described in more detail in the chapter of radiopharmaceutical and biokinetics as well as in the chapter clinical results. The quantitative analysis depends on the computer program used. Split functions representing uptake, excretion, and biliary retention should be calculated. Normally it is necessary to generate various ROIs at least a ROI of the whole liver without gallbladder and a peripheral ROI which should not be influenced by a hepatic bile. Some authors choose a ROI over the heart region to measure the disappearance of activity.

SPECT
SPECT is not in common practice in the differential diagnosis of jaundice. However it may be useful in the localization of intrahepatic focal activity retentions for instance in Caroli syndrome, after trauma, or in the diagnosis of a bilioma.

Clinical results

Intrahepatic vs. extrahepatic jaundice
Jaundice belongs to the key indications of hepatobiliary scintigraph. Jaundice can be classified as parenchymal or as obstructive jaundice. The latter one must be subdivided into incomplete and complete obstruction (Table 3).

It is important to differentiate the underlying mechanism of jaundice as early as possible, because of the life serving consequences in diagnosis and therapy, be it medical or surgical in nature. Reasons for hepatocellular jaundice are cirrhosis, different types of hepatitis, metabolic disorders for instance Rotor syndrome, and toxic cholestasis.

Sensitivity and specificity of HBS to differentiate between intra- and extrahepatic jaundice is about 90% [20, 29, 43, 54].

Toxic cholestasis
Although being due to a disorder of the bile capillaries the biokinetic abnormalities of cholestasis is predominantly a disease of the parenchyma. Toxic cholestasis can be characterized by a normal uptake function measured as

T_{max} or extraction fraction or – as we do – on account of half-life of uptake and an increase of transit time in a patient with activity excretion in the duodenum.

Mild subclinical toxic reactions of bile capillaries occur in association with hepatotoxic drugs. The course can be monitored by HBS [6].

Parenchymal jaundice

In contrast to toxic cholestasis primary hepatocellular disease is a pathophysiological combination of an impaired uptake as well as an increased time of parenchymal transit. Differentiation between a primary perfusion failure as in cirrhosis or a primary extraction failure as in hepatitis is not discernible by means of clinical methods.

In toxic cholestasis and hepatocellular disease activity in the duodenum is seen, but hepatic and common bile ducts in most of the cases remain invisible. Independent of the duodenal activity the biliary retention – which is a measure of the radioactivity pool in the greater bile ducts – is the second key function in obstruction. The possibility to measure this function directly as retention offers a significant assistance especially in case of a widely reduced excretion function where biliary ducts are not visualized.

Sensitivity of HBS in patients with an impaired parenchymal function is 90% exceeding the sensitivity of sonography [2, 12].

Complete vs. incomplete obstruction

Complete obstruction is potentially diagnosed by late images (24 h) presenting with complete lack of activity in the duodenum [20]. Incomplete obstruction is recognised by a positive gut scan. In most cases bile ducts can be visualized in incomplete obstruction [3], often as dilated structures (Fig. 3).

After successful treatment of partial obstructions biliary retention (RET_{40}) will be reduced within 1 week although bile ducts are still dilated in sonography or CT. This underscores the complementary roles of morphologic and functional methods for proper decision making.

Sensitivity and specificity. Sensitivity and specificity in patients with jaundice to differentiate mechanism of jaundice are: between 90 and 100% for incomplete obstruction, but only a few percent lower for complete obstruction.

Sonography and HBS in the diagnostic course

HBS is the second non invasive step after clinical investigation including tests of clinical pathology and ultrasound or CT [26]. It is a good policy to interpose HBS between ultrasound or CT. HBS has a high accuracy in differentiating intra- and extrahepatic jaundice [12, 14, 20, 53] and invasive procedures of interventional radiology which are not free of risk [25, 30, 34]. An incomplete obstruction without dilatation of bile ducts can only be diagnosed non invasively by HBS [35, 48, 55]. A further reason why HBS

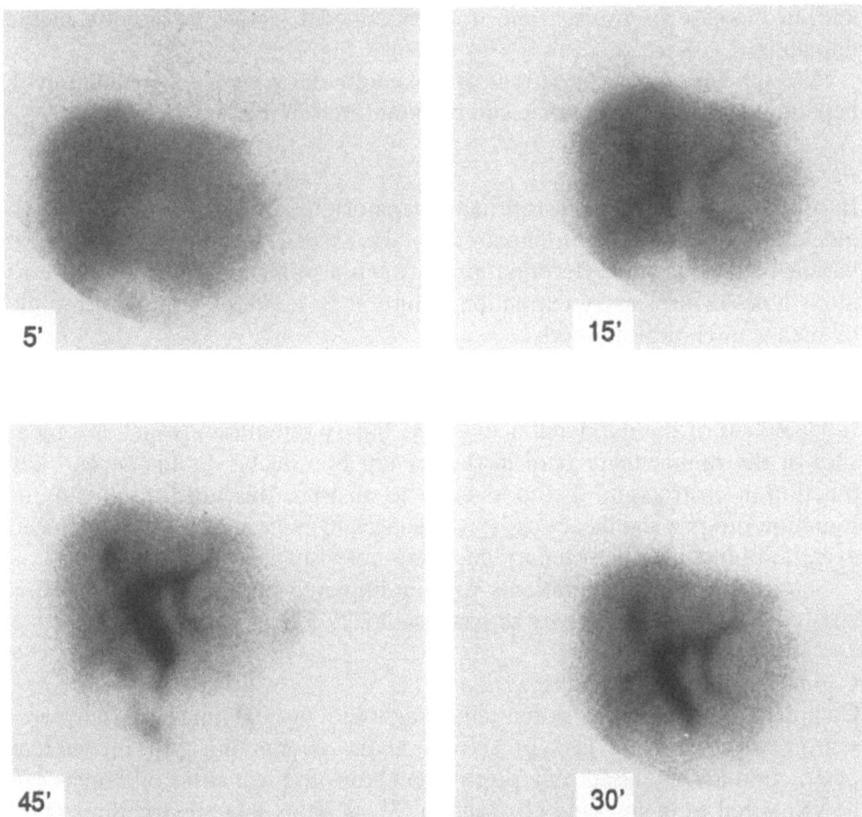

Figure 3. Hepatobiliary scintigraphy of a 80 years old female patient with an incomplete obstruction of the bile duct due to a tumour. Control after endoscopic implantation of a prosthesis in the bile duct. Elimination of the bile is still insufficient. Times and arrangement of the scintigrams are the same as in Fig. 1. HBS functional parameters are HLU: 2.8 min, MPTT: 91 min, RET_{40} 18%.

should be applied before invasive procedures is that after the localisation of the lesion invasive procedures can be applied more effectively by guiding to the localisation of obstruction more directly. Furthermore in cases where an obstruction can be excluded invasive procedures are not indicated.

Time after onset of jaundice

The effectivity of HBS depends on the time after the onset of jaundice [5, 20, 21]. In addition the risk of the patient increases with time independent of the applied diagnostic procedures [37]. Therefore it is important to use HBS immediately after onset of jaundice as a second non-invasive method in addition to ultrasound. The accuracy of HBS decreases with increasing bilirubin values [47]. If no sufficient uptake can be achieved within the first 2 h HBS should be terminated [26].

Multiparameter analysis

For this paper we reevaluated previous own data [7] with a linear discriminant analysis [1]. A total of 73 patients were investigated. 52 had a normal liver function due to history and clinical pathology. This patient group was investigated for a duodenal gastral reflux. 10 patients suffered from an incomplete obstruction and 12 from a cirrhosis. Using all 3 functions HUL, MPTT, and RET_{40} we could discriminate patients from healthy persons with a sensitivity of 90%, a specificity of 88%, and an accuracy of 89%. Nearly the same good result is achieved if we use only the MPTT and RET_{40} (Fig. 4). Cirrhosis and incomplete obstruction was differentiated with an accuracy of 95%. Theses results must be confirmed by a further group of patients. However, the data demonstrate that a multiparameter analysis is a powerful diagnostic tool to improve HBS.

Neonatal jaundice

Neonatal jaundice is not recognized on the basis of anatomical imaging alone. Ultrasound and other non-invasive procedures are frequently not conclusive. The main differential diagnosis in this case is neonatal hepatitis vs. atresia [13, 15]. The latter can be diagnosed very similarly according to the signs of a complete obstruction in adults. Activity in the gut excludes the total obstruction. This diagnosis can be made by qualitative evaluation without a loss of diagnostic accuracy. In cases with an incomplete obstruction activity in the duodenum is seen.

Today radiopharmaceuticals of choice are IDA-agents with fast excretion [13].

Postoperative jaundice

Surgical procedures of the bile duct system are often required. Therefore postoperative complaints are not so rare [28]. The main postoperative complication is a bile leakage which mostly does not induce jaundice.

Dilatation of bile ducts as the key sign of obstruction for ultrasound is not a useful tool in patients after surgery of the bile ducts or bile bladder because dilated biliary pathways are a frequent finding after surgery [40, 45, 59]. After extraction of gallbladder hepatic and common bile ducts dilate because they take a new function as a bile reservoir. Therefore HBS is the non invasive procedure of choice in postoperative jaundice [4, 28].

Reasons of postoperative jaundice. Persisting jaundice after surgery is a potential outcome. Other reasons are obstructions of the common bile duct after operative treatment [41, 57]. Further a bilioma, which may increase its size over years can be a reason for jaundice [58].

Especially in those cases with a preoperative jaundice it can be a problem

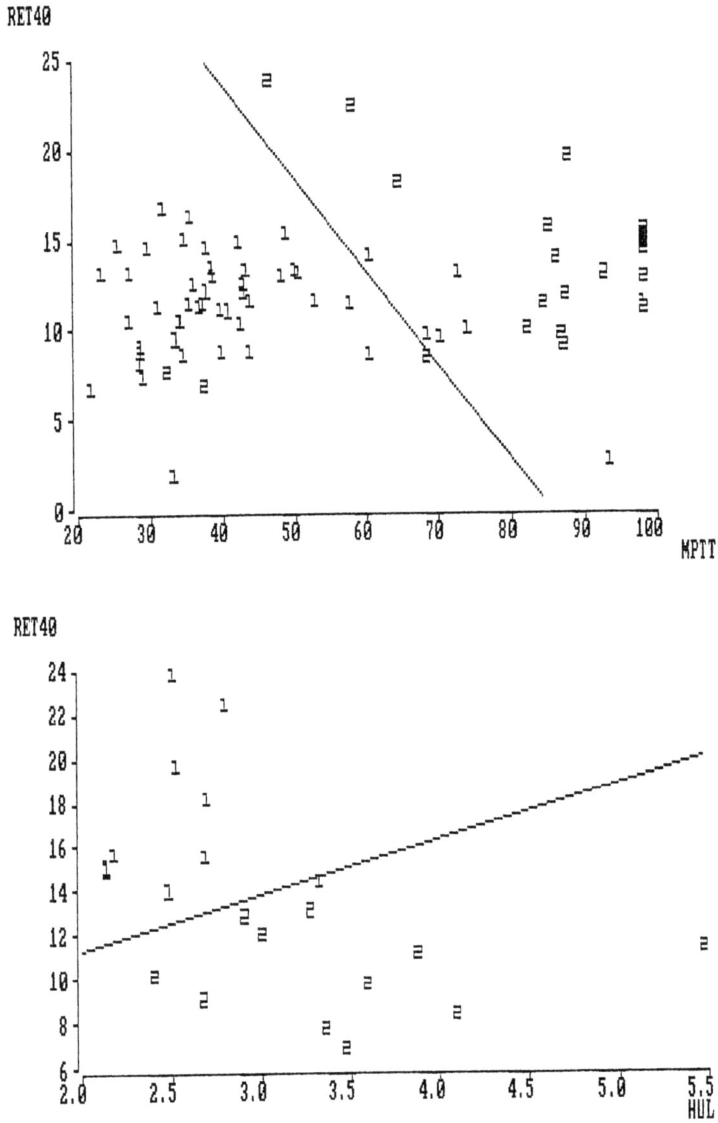

Figure 4. Two plots of discriminant analysis. Upper: discrimination of normals (group 1) vs. pathologic cases (cirrhosis or incomplete obstruction, group 2). Parameters are MPTT and RET_{40}. The accuracy is 86%. Lower: discrimination of obstructive jaundice (group 1) vs. parenchymal jaundice (cirrhosis, group 2). Parameters are HLU and RET_{40}. Accuracy is 89%.

to differentiate between a parenchymal and an obstructive disease. In all these instances HBS can be helpful. Diagnostic rules are not different from those before surgery. Sensitivity and specificity are in the range of intra- or posthepatic jaundice before operation [58].

Patients after trauma
Jaundice as the outcome of a trauma is also a rare situation. In some cases hematomas or traumatic cysts can be responsible for an obstruction. Particularly after an acute trauma a shock liver can be the reason of total functional loss.

Outlook and conclusions

In contrast to the diagnostic potentials of HBS its acceptance by clinicians is often low. One major reason probably is that gastroenterologists are prepared to use semi-invasive and invasive procedures instead of HBS very early in the diagnostic course.

Another reason is that many of the hepatocellular diseases can be diagnosed from the patients history and clinical pathology for instance by enzyme tests for all kinds of hepatitis.

Multiparameter analysis
On the other hand potentials should be acknowledged that the HBS deserves more attention by gastroenterologists. A major impact for HBS in future is the use of more standardized computer programs with the possibility of a multiparametric analysis of HBS.

As shown by Krishnamurty [27] and his coworkers as well as by our group it is possible to define functional topographic parameters as well as the combination of those parameters which enables us to differentiate between the major pathophysiological entities and influence the diagnostic course to make it quick, sure, and safe in patients with jaundice.

Conclusions

In conclusion (1) HBS is a useful method which helps to differentiate various functional mechanisms of jaundice including toxic reaction of the bile capillaries which are not visible in other morphological examinations. (2) It is mentionable that quantification improves HBS. (3) It is important to use HBS in the very early stage of a disease because of the fact that otherwise hepatobiliary function is depressed to a degree that no sufficient extraction can be achieved. (4) It helps to estimate the severity and quantify functional impairment in acute and chronic diseases of the hepatic parenchyma. For this question especially the normalised mean parenchymal transit time was helpful in our hands. (5) HBS is useful for follow up.

References

1. Ackermann H (1990) *Biometrische Analyse von Stichproben (BIAS)*, Programmversion 1.1, W.B. Verlag Winfried Bender.
2. Biersack HJ, Breuel H-P, Altland H, Bell E (1979) 'Morphologische und funktionelle Beurteilungskriterien der hapatobiliären Funktionsszintigraphie mit ^{99m}Tc-Diäthyl-IDA bei Parenchymikterus.' *Nucl Med* 18: 204.
3. Biersack HJ, Lindstaedt H, Thelen M et al. (1980) 'Hepatobiliäre Funktionsszintigraphie mit IDA-Derivaten – Ihre Bedeutung im Vergleich zur endoskopische retrograden Cholangiographie', in: Höfer R and Bergmann (Hrsg.). *Radioaktive Isotope in Klinik und Forschung*, Egermann, Wien, pp 99.
4. Biersack HJ (1982) 'Vergleichende Gegenüberstellung von hepatobiliärer Funktionsszintigraphie und Sonographie / CT / ERCP / PTC,' *Der Nuklearmediziner* 5: 39.
5. Blue PW (1985) 'Hyperacute complete common bile duct obstruction demonstrated with ^{99m}Tc-IDA cholescintigraphy.' *Nucl Med Commun* 6: 275.
6. Brandhorst I, Maul FD, Hör G (1985) 'Quantitative hepatobiliäre Funktionsszintigraphie bei Hauterkrankungen,' in Holzmann H, Altmeyer P, Hör G, Hahn K (Hrsg.). *Dermatologie und Nuklearmedizin* Springer Verlag. pp. 330
7. Brandhorst, I (1988) *Die quantitative hepatobiliäre Sequenszintigraphie Inauguraldissertation*, Frankfurt.
8. Brown PH, Juni JE, Liebermann DA et al. (1988) 'Hepatocyte vs. biliary disease: a distinction by deconvolution analysis of ^{99m}Tc IDA time-activity curve.' *J Nucl Med* 29: 623.
9. Chervu LR, Nunn AD, Loberg MD (1982) 'Radiopharmaceuticals for hepatobiliary imaging.' *Semin Nucl Med* 12: 5.
10. Collier BD, Treves S, Davis MA, Heyman S, Subramanian G, McAfee JG (1980) 'Simultaneous ^{99m}Tc-p-butyl-IDA and ^{131}I-rose-bengal-scintigraphy in neonatal jaundice.' *Radiology* 134: 719.
11. Delpart GD (1923) 'Studies on liver function: rose bengal elimination from blood as influenced by liver injury.' *Arch Intern Med* 32: 401.
12. Dewbury GD, Clark B (1979) 'The accuracy of ultrasound in the detection of cirrhosis of the liver.' *Br J Radiol* 52: 945.
13. Eissner D, Hahn K, Baumann W, Peters H (1982) 'Hepatobiliäre Sequenz- und Funktionsszintigraphie in der Pädiatrie.' *Der Nuklearmediziner* 1: 27.
14. Frank Th, Albers G, Voigt W, Kahaly G (1981) 'Bildgebende Diagnostik von Leber und Pankreas unter besonderer Berücksichtigung des Cholestase-Syndroms.' *Röntgenstr* 45: 4.
15. Gates GF, Sinatra FR, Thomas DW (1980) 'Cholestatic syndromes in infancy and childhood.' *Amer J Roentgenol* 134: 1141.
16. Harvey E, Loberg M, Cooper M (1975) '^{99m}Tc HIDA: A new radiopharmaceutical for hepatobiliary imaging.' *J Nucl Med* 16: 533.
17. Harvey E, Loberg M, Ryan J, Sikorski S, Faith M, Cooper M (1979) 'Hepatic clearance mechanism of ^{99m}Tc-HIDA and its effect on quantification of hepatobiliary function. Concise communication.' *J Nucl Med* 20: 310.
18. Hör G, Kempken K, Pabst HW, Maul FD (1981) 'Nuklearmedizinsche Gallenwegsdiagnostik.' *Therapiewoche* 31: 3.
19. Hottenrott C, Böttcher W, Maul, FD, Wildgrube HJ, Peters A (1983) 'Diagnostische transumbilikale Shunt-Simulation als Indikationshilfe zum portosystemischen Shunt.' *Chirurg* 54: 149.
20. Kempken (1982) 'Hepatobiliäre Sequenz- und Funktionsszintigraphie in der Differenzialdiagnose des Ikterus.' *Der Nuklearmediziner* 5: 11.
21. Klingensmith WC, Whitney WP, Spitzer VM et al. (1981) 'Effect of complete biliary obstruction on serial hepatobiliary imaging in an experimental model: Concise communication.' *J Nucl Med* 22: 866.

22. Klingensmith WC, Fritzberg AR, Spitzer VM, Kuni CC, Shanahen WSM (1981) Clinical comparison of diisopropyl-IDA-^{99m}Tc and diethy-IDA ^{99m}Tc for evaluation of the hepatobiliary system.' *Radiology* 140: 791.
23. Klingensmith WC, Kuni CC, Fritzberg AR (1982) 'Cholescintigraphy in extrahepatic biliary obstruction.' *AJR* 139: 65.
24. Klingensmith WC, Fritzberg AR, Spitzer VM, Kuni CC, Williamson MR, Gerhold JP (1983) 'Work in progress: clinical evaluation of ^{99m}Tc-trimethy-bromo-IDA for hepatobiliary imaging.' *Radiology* 146: 181.
25. Kreek MJ, Balint JA (1980) '"Skinning needle" cholangiography-results of a pilot study of a voluntary prespective method for gathering risk data on new procedures.' *Gastroenterology* 78: 598.
26. Krishnamurthy GT, Krishnamurty (1988) 'Scintigraphic criteria for the diagnosis of obstructive hepatobiliary disease with ^{99m}Tc IDA.' *Clin Nucl Med* 13: 704.
27. Krishnamurthy GT, Turner FE (1990) 'Pharmacokinetics and clinical application of ^{99m}Tc-labeled hepatobiliary agents.' *Semin Nucl Med* 20: 130.
28. Kune GA (1972) *Current Practice of Biliary Surgery*. Boston: Little, Brown and Co.
29. Lee A, Ram MD, Shih WJ, Murphy K (1980) '^{99m}Tc BIDA biliary scintigraphy in the evaluation of jaundiced patients.' *J Nucl Med* 17: 1407.
30. Liebermann DA, Krishnamurty GT (1986) 'Intrahepatic vs. extrahepatic cholestasis discrimination with biliary scintigraphy combined with ultrasound.' *Gastroenterology* 90: 734.
31. Loberg MD, Callery PS, Harvey E, Faith W, Cooper M (1975) Development of a chelating group for synthesis into drug and biological analogues.' *J Nucl Med* 16: 546.
32. Loberg MD, Cooper M, Harvey E, Callery P, Faith W (1976) 'Development of new radiopharmaceutical based on N-substitution of iminodiacetic acid.' *J Nucl Med* 17: 633.
33. Ludovici M (1985) 'Quantitative hepatobiliäre Funktionsszintigraphie mit Diäthyl-IDA: Biokinetische Untersuchungen zur Bedeutung der Parameter eines halbautomatischen Auswerteprogrammes.' *Inaugural-Dissertation*, Frankfurt.
34. Matzen P, Malchow-Moller A, Brun B, et al. (1983) 'Ultrasonography, comuted tomography and cholescintigraphy in suspected obstructive jaundice – a prospective comparative study.' *Gastroenterology* 84: 1492.
35. Maul FD, Standke R, Brandhorst I, Eggert UE, Jessen K, Hör G (1982) 'Quantitative hepatobiliäre Funktionsszintigraphie in der nuklearmedizinischen Praxis.' *Der Nuklearmediziner* 5: 95.
36. Maul FD, Bittner G, Baum RP, Hör G, Hottenrott C, Brandhorst I, Standke R (1988) 'Functional pattern of hepatobiliary split function after liver transplantation: a new approach on a biokinetic analysis.' *Eur J Nucl Med* 14: 238.
37. Middendorp UG (1979) *Klinische Aspekte des Verschluß-Ikterus*. Bern Stuttgart Wien: Huber.
38. Miller JH, Sinatra FR, Thomas DW (1980) 'Biliary excretion disorder in infants: Evaluation using ^{99m}Tc-PIPIDA.' *Amer J Roentgenol* 135: 47.
39. Nunn AD, Loberg MD, Conley RA (1983) 'A structure-distribution-relationship approach leading to the development of ^{99m}Tc mebrophenin: an improved cholescintigraphic agent.' *J Nucl Med* 24: 423.
40. Oddi R (1887) D'une disposition a sphincter specale de l'ouverture du canal choledoque Arch.' *Biol* 8: 317.
41. Reichelt HG (1982) Die Bedeutung der hepatobiliaren Sequenz- (Funktions) Szintigraphie (HBS(F)S) in der Erfolgsbeurteilung und Abklärung von Symptomen nach operativen Eingriffen an Leber und Gallenwegen.' *Nuklearmed* 5: 69.
42. Reichelt HG, Popescu HI (1979) 'The importance of liver uptake and retention indices in assessment of clinical usefulness of hepatobiliary imaging agents.' *J Nucl Med* 20: 171.
43. Pauwels S, Piret L, Schoutens A, Vandermoten G, Beckers L (1980) ^{99m}Tc-diethy-IDA imaging: Clinical evaluation in jaundice patients.' *J Nucl Med* 21: 1022.
44. Porter DW, Loberg MD, Eacho PI, Weiner M (1979) 'Comparison of hepatobiliary radiopharmaceuticals in an in-vitro-model.' *J Nucl Med* 20: 642.

45. Rettenmaier G, Seitz KH (1977) 'Ultraschalluntersuchungen beim Ikterus.' *Dtsch med Wschr* 102: 1565.
46. Rosenthal L (1982) 'Cholescintigraphy in presence of jaundice utilizing ^{99m}Tc-IDA.' *Semin Nucl Med* 12: 53.
47. Rosenthal L, Damtew B, Kloiber R, Warshawski R (1983) 'The difficulty of estimating biliary tract distention by radionuclide imaging.' *Diagn Imaging* 50: 154.
48. Ryan J, Isikoff M, Nagle C, Loberg M, Buddemeyer E, Cooper M (1976) 'The combined use of ^{99m}Tc-HIDA and ultrasound in differential diagnosis of jaundice.' *J Nucl Med* 17: 545.
49. Schwarzrock R, Kotzerke J, Hundeshagen H, Böcker K (1986) ^{99m}Tc-diethyl-iodo-HIDA (IODIDA): A new hepatobiliary agent in clinical comparison with ^{99m}Tc-diisopropyl-HIDA (DISIDA) in jaundiced patients.' *Eur J Nucl Med* 12: 346.
50. Spitz J, Hildebrandt N, Clemenz N, Schattenberg J, Weigand H (1987) 'Klinische Relevanz und diagnostische Aussagekraft von ^{99m}Tc-Diäthy-Iod-IDA (IOD IDA) bei Patienten mit erhöhten Bilirubin-Spiegeln im Vergleich zu ^{99m}Tc-Diäthy-IDA (Hepatobida).' *Nuc Compact* 18: 61.
51. Taplin GV, Meredith OM, Kade H. (1955) 'The radioactive (^{131}I tagged) rose-bengal uptake excretion test for liver function using external gamma ray scintillation counting techniques.' *J Lab Clin Med* 45: 665.
52. Tarolo GL, Picossi R, Palagi B, Cammelli F (1981) 'Comparative quantitative evaluation of hepatic clearance of diethyl-IDA and para-butyl-IDA in jaundiced and non-jaundiced patients.' *Eur J Nucl Med* 6: 539.
53. Taylor KJW, Rosenfield AT (1977) 'Grey scale ultrasonography in the differential diagnosis of jaundice.' *Arch Surg* 112: 820.
54. Weissmann HS, Rosenblatt R, Sugarman LA et al. (1980) 'The role of nuclear imaging in evaluating choestasis-an update.' *Semin Ultrasound* 1: 134.
55. Weissmann HS, Rosenblatt R, Sugarman LA, Badia JD, Freeman LM (1980) 'Early diagnosis of acute common bile duct obstruction by ^{99m}Tc-IDA (iminodiacetic acid) cholescintigraphy.' *J Nucl Med* 21: P 41.
56. Weissmann HS, Sugarman LA, Freeman LH (1981) 'The clinical role of ^{99m}Tc iminodiacetic acid cholescintigraphy,' in: Freeman LM, Weissmann HS (eds), *Nuclear Medicine Annual*, p. 35. New York: Raven Press.
57. Weissmann HS, Gliedman ML, Wick PJ, Sugarman LA, Badia J, Guglielmo K, Freeman LM (1982) 'Evaluation of the postoperative patient with ^{99m}Tc-IDA cholescintigraphy.' *Semin Nucl Med* 12: 27.
58. Weissmann (1983) 'Role of ^{99m}Tc-IDA scintigraphy in the evaluation of hepatobiliary trauma.' *Semin Nucl Med* 13: 199.
59. Wise RE (1968) *Intravenous choleangiography*. Springfield, Illinois: Charles Co. Thomas.
60. Wistow BW, Subramanian G, van Hertum RL, Henderson RW, Gayne GM, Hall RC, McAfee JG (1977) 'An evaluation of ^{99m}Tc-labeled hepatobiliary agents.' *J Nucl Med* 00: 455.
61. Wistow BW, Subramanian G, Gagne GM, Herdessen RW, McAfee JG, Hall RC, Grossman ZD (1978) 'Experimental and clinical trials of new ^{99m}Tc-labeled biliary agents.' *Radiology* 128: 793.
62. Wolf F, Krönert E (1978) 'Leber und Gallenwege,' in: Diethelm L, Heuck F, Olsen O, Strnad F, Vieten H, Zuppinger A (eds), *Handbuch der medizinischen Radiologie*, Band 15/2.

3. Kinetics of gallbladder emptying

ASLAM R. SIDDIQUI and HENRY N. WELLMAN

Physiology

The main function of the gallbladder is to store and concentrate the bile. The concentration, which can be as much as tenfold, occurs by reabsorption of the water and electrolytes (mainly sodium, bicarbonate and chlorides). The storage of the bile probably is helped by tonic contraction of the sphincter of Oddi, which results in a positive pressure of 3 to 10 mm of mercury in the common hepatic duct of an adult.

The emptying of the gallbladder is achieved by the contraction of the smooth muscles in its walls and the relaxation of the ampullary sphincter. Hormonal and, to a much lesser extent, neurogenic mechanisms are involved in the gallbladder emptying. Studies in animals have shown that the gallbladder relaxes with the sympathetic nervous system stimulation and contracts with the parasympathetic stimulation [1]. In humans, there are conflicting reports about the role of nervous mechanism on the gallbladder contractility. It appears that, in general, vagotomy does not influence the gallbladder response to cholecystokinin [2].

Hormones have the most important role in the determination of gallbladder motility. Traditionally, cholecystokinin (CCK) has been regarded as the main hormone in stimulating the gallbladder contraction, however, there are other hormones which have cholecystokinetic properties also. These are motilin, gastrin I, gastrin II and cerulein (derived from the skin of an Australian frog and not found in humans). CCK is released from the mucosa of the small bowel in response to the presence of fatty acids and amino acids in the duodenum. It acts directly on the smooth muscles of the gallbladder to cause contraction and also relaxation of the ampullary sphincter. The muscle contraction most likely is through the stimulation of intracellular cyclic 3′, 5′-guanosine monophosphate [2].

Nuclear medicine evaluation

Gallbladder emptying after a fatty meal or administration of synthetic CCK can be evaluated by contrast cholecystography, ultrasonography or ma-

H.J. Biersack and P.H. Cox (eds), Nuclear Medicine in Gasteroenterology, 37–46

nometry [3, 4, 5]. These techniques have certain limitations. Some are invasive, while others require use of contrast agents, which can cause alterations in physiology and occasionally are associated with undesirable side effects. Most importantly, the majority of these tests do not provide any objective data. The radionuclide studies, on the other hand, do not require sedation, intubation, or large quantities of contrast media and they provide quantitative information.

The use of non-invasive nuclear medicine techniques to study the gallbladder physiology (contraction and emptying) quantitatively is not new. Iodine-131 labeled iopanoic acid was the first hepatobiliary agent used for this purpose [6, 7]. The counts over the liver were monitored by an external detector. The lack of morphological details made the tests somewhat unreliable.

In the mid-1970's technetium-99m labeled N-substituted iminodiactic acid (Tc-IDA) derivatives were introduced as hepatobiliary imaging agents [8]. The advantages of these compounds are rapid blood clearance, little renal excretion and ease of labeling. The biodistribution and the biliary kinetics of various IDA derivatives have been studied extensively, however, there does not appear to be a consensus regarding the agent of choice [9].

There are several reported methods of imaging the gallbladder for the assessment of its function. Here we describe a basic protocol which is simple and can be modified to suit the needs of individual institutions. All patients should fast for at least two hours, preferably four hours, prior to the administration of Tc-IDA. In one study of normal volunteers, the visualization of gallbladder was reduced from 100% in the fasting stage to 36% in the post prandial state [10]. Endogenous CCK, which is released in response to food, causes the gallbladder to contract and empty with a half-time of 45 min. This results in marked reduction in the net flow of the bile and the radiotracer into the gallbladder for a few hours after meals. Fasting for more than 24 h also is not desirable. The nonvisualization of the gallbladder in this situation is most likely due to the distention and bile stasis. Patients on hyperalimentation, or with a history of fasting for more than 24 h, should receive synthetic CCK approximately 30 min before the injection of the radiotracer [11].

The usual adult dose of Tc-IDA is 185 MBq(5 mCi). After intravenous injection, sequential images of the liver are obtained using a gamma camera equipped with a general purpose collimator and interfaced to a computer. At 60 minutes, or when the gallbladder is maximally filled (little or no activity in the liver), 0.02 μg/kg of CCK is injected intravenous slowly over 3 min. Following this, four to six anterior images at 5 min intervals are obtained for a preset time. The computer acquisition is started one minute prior to the injection of CCK and continued until the end of the study. The gallbladder emptying is quantified by measuring the ejection fraction. The ejection fraction is determined by creating a region of interest over the gallbladder and an equal one over an adjacent part of the liver (background) on baseline and all post-CCK images. The total number of pixels and counts from each region are obtained and the gallbladder ejection fraction is evaluated according to the following formula [12]:

$$\frac{(\text{Net pre CCK gallbladder cts}) - (\text{Net post CCK } (t) \text{ GB cts})}{(\text{Net pre CCK gallbladder cts})} \times 100 = \%\text{EF}$$

where t is the time post-CCK administration and the net gallbladder counts = (total gallbladder counts) – (backgrcund counts/pixel × number of gallbladder pixels).

Some variations of the technique are use of the pin-hole collimator or imaging in the right lateral projection to eliminate superimposed radioactivity in the duodenum. Fatty meals and continuous infusion of OCK have also been used for the measurement of gallbladder emptying [13, 14].

Normal values

We use 35% as the lower limit of normal for gallbladder ejection fraction, mean being approximately 59% [15]. Doubling the dose of CCK (0.04 mg/kg) does not alter the ejection fraction [16], however, the refilling of the gallbladder which probably reflects the recovery of sphincter tone, is delayed [17]. The gallbladder starts emptying 2 min after the injection of CCK and this lasts for approximately 11 min; the mean ejection rate is approximately 6% per minute. The ejection fraction is somewhat higher in the females, however, there is no difference in the gallbladder emptying rates in the male and female [14, 18]. The age of the patient does not influence the ejection fraction or the emptying rate [19].

Some other quantitative data about the gallbladder function can be obtained during this study. An examination of the time-activity curves generated from the gallbladder region after the injection of the radiotracer can reveal peak filling rate, which is the maximum rate of increase in the counts from minute to minute (normal = 30 min after the injection of the radiotracer) and the time to maximum filling of the gallbladder (normal = 63 min) [16].

Acalculous gallbladder disease

Since the indications for cholecystectomy, in the patients with gallstones, are almost universally accepted, the main impact of this diagnostic technique has been in the management of the patients with acalculous biliary disease. Acalculous cholecystitis can be acute or chronic and makes up about 10%

of all cholecystitis; the incidence is probably increasing [20, 21]. Acute calculous and acalculous cholecystitis are indistinguishable from each other clinically, pathologically and in their surgical management. The hepatobiliary imaging findings are also similar in these two conditions, i.e. nonvisualization of gallbladder [22].

Substances toxic to the gallbladder wall are considered to be important in the genesis of chronic acalculous cholecystitis. These are pancreatic enzymes and hyperconcentrated bile [23]. The reflux of pancreatic juices into the gallbladder occurs as a result of pancreatitis and/or surgically altered anatomy. Hyperconcentration of the bile usually is secondary to increased pigments (such as in anemia or after multiple transfusions) or biliary stasis. The biliary stasis is due to altered dynamics (e.g., volume depletion, hyperalimentation, etc.) or cystic duct obstruction from mucosal edema, fibrosis, kinking or extrinsic pressure on the duct. The non-calculous cystic duct obstruction is also known as cystic duct syndrome [24].

The patients with chronic acalculous cholecystitis present with dyspeptic symptoms (fatty food intolerance, flatulence, epigastric fullness, upper abdominal discomfort, heartburn, etc.) and present a challenging, as well as, frustrating diagnostic and therapeutic problem. The usual diagnostic tests such as oral cholecystogram, ultrasonography and hepatobiliary imaging are normal. CCK cholecystography has been of some value and is considered diagnostic when the following criteria are met: (1) duplication of the patients symptoms after the injection of CCK, (2) failure of the gallbladder to empty, and (3) abnormal gallbladder contraction [25]. However, due to the subjective nature of the criteria, this examination is not as reliable or useful as was originally suggested [26].

The quantitative assessment of the gallbladder function, as determined by measuring the ejection fraction, is an accurate nuclear medicine test in patients with acalculous biliary disease. Among the patients with histologic changes of chronic cholecystitis, hepatobiliary scintigraphy (gallbladder ejection fraction less than 40%) has an approximately 90% sensitivity compared with 60 to 65% for sonography and oral cholecystography [12, 27, 28]. There is some controversy regarding the identification of the patients who may obtain long-term symptomatic relief after cholecystectomy. Some studies suggest that gallbladder ejection fraction measurements fail to identify suitable surgical candidates [27, 29], whereas others have shown consistent symptomatic relief after cholecystectomy in patients with low ejection fractions [12, 30]. In one study, approximately 25% of the gallbladders removed from the patients with low gallbladder ejection fractions were normal clinically and histologically [31]. These patients may have had cystic duct abnormalities. There seems to be no consensus or statistical information, still it appears to us from the general trend of the literature and our own unpublished work, that most symptomatic patients with low gallbladder ejection fractions do benefit from cholecystectomy, though the placebo effect of the surgery should not be underestimated.

Calculous gallbladder disease

Abnormal gallbladder motor function has been postulated to be a factor in the development of gallstones [32, 33]. The stasis of bile provides the time necessary for the precipitation of cholesterol microcrystals which then grow to macroscopic stones. This sequence of events has been documented in experimental animals [34]. The gallbladder function in humans with gallstones can be assessed with quantitative biliary scintigraphy.

The gallbladder filling is normal in these patients [15, 18, 35, 36], however, in response to a fatty meal the ejection fraction is consistently and significantly decreased when compared to the normals [18, 35]. The gallbladder ejection fractions measured after the administration of CCK are variable. In general, the gallbladder ejection fraction is reduced [15], however, in one study two distinct groups were identifiable – one with marked reduction and a second with minimal reduction [36]. Fisher et al. [19] have reported normal CCK gallbladder ejection fractions in patients with stones and Masclee et al. [37], in fact, have shown an accelerated gallbladder emptying. No discriminatory factor could be found in these groups of patients and the clinical significance of this finding is unknown.

Gallstones are more common in females, suggesting a relationship between gender and gallbladder function. Decreased gallbladder emptying is seen during the progesterone peak of the menstrual cycle and during the last two trimesters of pregnancy [38]. Progesterone is a known smooth muscle relaxant and has been found to have receptors in the gallbladder wall. In one study, 60% of the patients with gallstones had progesterone receptors [39]. The patients with receptors had significantly lower gallbladder ejection fractions when compared to the patients without receptors. The percentage of normal controls who are progesterone receptor-positive is unknown.

There is no significant adherence of the IDA radiopharmaceutical to the gallstones and it does not absorb or adsorb in the gallbladder wall [36], therefore, this cannot be the explanation of the finding of low ejection fraction in patients with gallstones. The gallbladders with stones are less distensible, but it is unlikely that this is responsible for a diminished contractile response to the fatty meal or CCK since the total counts over the gallbladder and the time to achieve these are not different for normals and patients with gallstones [40]. The reduction in gallbladder emptying is not secondary to an abnormality of gastric emptying, since these patients have normal gastric emptying [19]. Controlled studies have shown that the age and the sex of the patients are not responsible for the different gallbladder ejection fractions in normals and in patients with gallstones [19]. In conclusion, it is clear that the abnormal gallbladder emptying in patients with gallstones is a functional disorder, but because of the variable responses to the fatty meal and CCK, it is not clear whether these functional changes are the cause or the effect of gallstones [35].

Bile salts (chenodeoxycholic acid and ursodeoxycholic acid) have been proven to be efficacious in the dissolution of cholesterol gallstones [41]. During the therapy, these agents significantly decrease the gallbladder filling and ejection fraction. The dissolution of the gallstones improves the emptying which is further enhanced when the bile salts are discontinued [42]. The gallbladders of the patients whose stones fail to dissolve show long-term filling and emptying abnormalities. Both ursodeoxycholic acid and chenodeoxycholic acid have the same effect on the gallbladder function.

A report on the effects of extracorporeal shock-wave lithotripsy on the gallbladder emptying, using ultrasonography, showed no immdiate or long-term adverse effects on the gallbladder motility [43]. These patients started out with less than normal emptying of the gallbladder in response to CCK and this abnormality was not abolished after removal of the stone. We have no nuclear medicine data on these patients.

Other conditions

Diabetes mellitus is considered a predisposing factor for cholesterol gallstones [44]; the exact mechanism remains unknown. The gallbladder ejection fraction is lower in the diabetics compared to normal controls; the gallbladder filling remains normal [45]. There is no association between peripheral neuropathy and impaired gallbladder emptying; however, the diabetics with autonomic neuropathy have significantly lower gallbladder ejection fractions than those without autonomic neuropathy [46]. This suggests that the gallbladder function is affected in the same way as the function of the smooth muscles of the blood vessels, urinary system and other parts of the gastrointestinal system.

Impaired gallbladder function has been reported in obese patients with and without diabetes mellitus. The compliance of the gallbladder is increased resulting in larger than normal volume. The gallbladder filling is rapid, but the emptying in response to both fatty meal and CCK is reduced when compared to the normal controls [46, 47]. It is unknown whether this demonstrable bile stasis plays a role in the increased incidence of gallstones in the obese patients.

Approximately 50% of the patients with sickle hemoglobinopathy develop pigmented gallstones by the age of 20 years [48]. In an ultrasound study of these patients, the fasting gallbladder volume, as well as the volume after a fatty meal ingestion was found to be greater than controls; however, the ejection fraction was normal [49]. In a small group of children with sickle cell anemia, we have seen decreased gallbladder ejection fractions in response to CCK. Perhaps bile stasis or retention in the gallbladder does have a role in the pathogenesis of gallstones in these patients.

Somatostatinomas are rare tumors and almost always are associated with gallstones [50]. In normal volunteers, administration of somatostatin markedly reduces the response of the gallbladder to CCK [51]. This suggests that the inhibition of gallbladder emptying by somatostatin may be an important factor in the development of gallstones in patients with somatostatinomas.

Adenomyomatosis of the gallbladder is a benign noninflammatory condition of uncertain etiology characterized by mucosal proliferation, muscle wall thickening and intramural diverticula [52, 53]. This entity is usually an incidental finding in a surgical specimen and the patients are asymptomatic. Occasionally, this condition may cause recurrent right upper quadrant pain. On hepatobiliary imaging, adenomyomatosis may cause focal wall motion abnormalities with or without a reduction in the gallbladder ejection fraction [53]. Theoretically, other conditions that can produce similar scintigraphic findings are compartmentalized gallbladder [54] and gallbladder duplication [55].

The measurement of the gallbladder ejection fraction using the nuclear medicine technique has shown that patients with severe pancreatic insufficiency have impaired gallbladder emptying after the administration of intraduodenal fat. This abnormality can be corrected by giving the patients pancreatic enzymes [56]. The impaired gallbladder emptying appears to be the result of a reduction in the release of CCK and it appears that the intraintestinal pancreatic enzymes have an important role in the intestinal phase of CCK secretion.

Conclusion

In summary, the dynamic function of the gallbladder can be assessed objectively and noninvasively following the injection of CCK or ingestion of fatty meal. The radionuclide method is count-based, therefore, the shape and the location of the gallbladder does not affect the test. Since almost every nuclear medicine facility is equipped with a gamma camera interfaced to a computer now, these studies can be performed universally. The information obtained is clinically relevant and most gastroenterologists and surgeons make treatment decisions based on the information the quantitative biliary scintigraphy provides.

References

1. Lundgrew O, Svannik J, Jivegard L (1989) 'Enteric nervous system. II. Physiology and pathophysiology of the gallbladder.' *Dig Dis Sci* 34: 284–288.
2. Banfield WJ (1975) 'Physiology of the gallbladder.' *Gastroenterology* 69: 770–777.
3. Hopman WPM, Rosenburch G, Jansen JBM, de Jong AJL, Lamers CBHW (1986) 'Gallbladder contraction: effect of fatty meals and cholecystkinin.' *Radiology* 157: 37–39.
4. Goldberg HI (1976) 'Cholecystokinin cholecystography.' *Semin Roentgenol* 11: 175–179.
5. Kishk SMA, Darweesh RMA, Dodds WJ, Lawson TL, Stewart ET, Kern MK, Hassanein EH (1987) 'Sonographic evaluation of resting gallbladder volume and post prandial emptying in patients with gallstones.' *AJR* 148: 875–879.

6. Englert E, Jr, Chiu VSW (1966) 'Quantitative analysis of biliary evacuation with a radioisotope technique.' *Gastroenterology* 50: 506–518.
7. Chapple MJ, Nolan DJ, Low-Beer TS, Davies ER (1975) 'Gallbladder emptying measured by radioisotope methods.' *Br J Radiol* 48: 19–22.
8. Loberg MD, Cooper M, Harvey E, Callery P, Faith W (1976) 'Development of new radiopharmaceuticals based on N-substitution of iminodiactic acid.' *J Nucl Med* 17: 633–638.
9. Krishnamurthy S, Krishnamurthy GT (1989) 'Technetium-99m-iminodiactic acid organic anions: review of biokinetics and clinical application in hepatology.' *Hepatology* 9: 139–153.
10. Klingensmith WC III, Spitzer VM, Fritzberg AR, Kuni CC (1981) 'The normal fasting and postprandial diisopropyl-IDA Tc99m hepatobiliary study.' *Radiology* 141: 771–776.
11. Larsen MJ, Klingensmith WC III, Kuni CC (1982) 'Radionuclide hepatobiliary imaging: nonvisualization of gallbladder secondary to prolonged fasting.' *J Nucl Med* 23: 1003–1005.
12. Fink-Bennett D, DeRidder P, Kolozsi W, Gordon R, Rapp J (1985) 'Cholecystokinin cholescintigraphic findings in the cystic duct syndrome.' *J Nucl Med* 26: 1123–1128.
13. Maton PN, Selden AC, Fitzpatrick ML, Chadwick VS (1984) 'Infusion of cholecystokinin octapeptide in men: relation between plasma cholecystokinin concentrations and gallbladder emptying rates.' *Eur J Clin Invest* 14: 37–41.
14. Mackie CR, Baxter JN, Grimme JS, Hulks G, Cuschieri A (1987) 'Gallbladder emptying in normal subjects – a data base for clinical cholescintigraphy.' *Gut* 28: 137–141.
15. Krishnamurthy GT, Bobba VR, McConnell D, Turner F, Mesgarzadeh M, Klingston E (1983) 'Quantitative biliary dynamics: introduction of a new noninvasive scintigraphic technique.' *J Nucl Med* 24: 217–223.
16. Sarva RP, Shreiner DP, Van Thiel D, Yingvorapant N (1985) 'Gallbladder function: methods for measuring filling and emptying.' *J Nucl Med* 26: 140–144.
17. Mesgarzadeh M, Krishnamurthy GT, Bobba VR, Langrell K (1983) 'Filling, postcholecystokinin emptying and refilling of normal gallbladder: effects of two different doses of CCK or refilling: concise communication.' *J Nucl Med* 24: 666–671.
18. Spellman SJ, Shaffer EA, Rosenthall L (1979) 'Gallbladder emptying in response to cholecystokinin: cholescintigraphic study.' *Gastroenterology* 77: 115–120.
19. Fisher RS, Stelzer F, Rock E, Malmud LS (1982) 'Abnormal gallbladder emptying in patients with gallstones.' *Dig Dis Sci* 27: 1019–1024.
20. Glenn F, Becker CG (1982) 'Acute acalculous cholecystitis: an increasing entity.' *Ann Surg* 195: 131–136.
21. Frykberg ER, Duong TC, LaRosa JJ, Etienne HB. (1988) 'Chronic acalculous gallbladder disease: a clinical variant.' *South Med J* 81: 1353–1357.
22. Swayne LC (1986) 'Acute acalculous cholecystitis: sensitivity in detection using technetium-99m iminodiactic acid cholescintigraphy.' *Radiology* 160: 33–38.
23. Munster AM, Brown JR (1967) 'Acalculous cholecystitis.' *Am J Surg* 113: 730–734.
24. Camishion RC, Goldstein F (1967) 'Partial noncalculous cystic duct obstruction (cystic duct syndrome).' *Surg Clin North Am* 47: 1107–1114.
25. Nora PF, McCarthy W, Sanej N (1974) 'Cholecystokinin cholecystography in acalculous gallbladder disease.' *Arch Surg* 108: 507–512.
26. Dunn FH, Christensen EC, Reynolds J, Jones V, Fordtran JS (1974) 'Cholecystokinin cholecystography. Controlled evaluation in the diagnosis and management of patients with possible acalculous gallbladder disease.' *JAMA* 228: 997–1003.
27. Raptopoulos V, Compton CC, Doherty P, Smith EH, D'Orai CJ, Patwardhan NA, Goldberg R (1986) 'Chronic acalculous gallbladder disease: multi-imaging evaluation with clinical-pathological correlation.' *AJR* 147: 721–724.
28. Realini S, Reiner M, Pescia R (1987) 'Study of gallbladder emptying using ^{99m}Tc-HIDA in acalculous cholecystopathy.' *Schweiz Med Wsch* 117: 1217–1220.
29. Sunderland GT, Carter DC (1988) 'Clinical application of the cholecystokinin provocation test.' *Br J Surg* 75: 444–449.
30. Pickleman J, Peiss RL, Henkin R, Salo B, Nagel P (1985) 'The role of sincalide cholescinti-

graphy in the evaluation of patients with acalculous gallbladder disease.' *Arch Surg* 120: 693–697.
31. Freeman IM, Sugarman LA, Weissmann HS (1981) 'Role of cholecystokinetic agents in ^{99m}Tc-IDA choscintigraphy.' *Semin Nucl Med* 11: 186–193.
32. Shaffer EA, Small DM (1976) 'Gallstone disease: pathogenesis and management.' *Curr Probl Surg* 13: 1–72.
33. Hoffman AF (1988) 'Pathogensis of cholesterol gallstones.' *Am J Gastroenterol* 10 (Suppl 2): S1–S11.
34. Fridhandler TM, Davison JS, Shaffer EA (1983) 'Defective gallbladder contractility in the ground squirrel and prairie dog during the early stages of the cholesterol gallstone formation.' *Gastroenterology* 85: 830–836.
35. Bobba VR, Krishnamurthy GT, Klingston E, Turner FE, Brown PH, Langrell K (1984) 'Gallbladder dynamics induced by a fatty meal in normal subjects and patients with gallstones: concise communication.' *J Nucl Med* 25: 21–24.
36. Pomeranz IS, Shaffer EA (1985) 'Abnormal gallbladder emptying in a subgroup of patients with gallstones.' *Gastroenterology* 88: 787–791.
37. Masclee AAM, Jansen JBMJ, Driessen WMM, Geuskens IM, Lamers CBW (1989) 'Plasma cholecystokinin and gallbladder responses to intraduodenal fat in gallstone patients.' *Dig Dis Sci* 34: 353–359.
38. Everson TE, McKinley C, Lawson M, Johnson M, Kern F (1982) 'Gallbladder function in the human female: effect of the ovulatory cycle, pregnancy, and contraceptive steroids.' *Gastroenterology* 82: 711–719.
39. Daignault PG, Fazekas AG, Rosenthall L, Fried GM (1988) 'Relationship between gallbladder contraction and progesterone receptors in patients with gallstones.' *Am J Surg* 155: 147–151.
40. Shaffer EA, McOrmand P, Duggan H (1980) 'Quantitative cholescintigraphy assessment of gallbladder filling and emptying and duodenogastric reflux.' *Gastroenterology* 79: 899–906.
41. Erlinger S, Go AI, Husson JM, Fevery J (1984) 'Franco-Belgian cooperative study of ursodeoxycholic acid in the medical dissolution of gallstones: a double-blind, randomized, dose-response study, and comparison with chenodeoxycholic acid.' *Hepatology* 4: 308–314.
42. Sylvestrowicz TA, Shaffer EA (1988) 'Gallbladder function during gallstone dissolution: effect of bile acid therapy in patients with gallstones.' *Gastroenterology* 95: 740–748.
43. Spengler V, Sackmann M, Sauerbruch T, Holl J. Paumgartner G (1989) 'Gallbladder mobility before and after extracorporcal shock-wave lithotripsy.' *Gastroenterology* 96: 860–863.
44. Lieber MM (1985) 'The incidence of gallstones and their correlation with other diseases.' *Ann Surg* 135: 394–404.
45. Stone BG, Gavaler JS, Belle SH, Shreiner DP, Peleman RR, Sarva RP, Yingvorapant N, Van Thiel DH (1988) 'Impairment of gallbladder emptying in diabetes mellitus.' *Gastroenterology* 95: 170–176.
46. Vezina WC, Paradis RL, Grace DM, Zimmer RA, Hutton LC, Chamberlain GW, Rycoft KM, Chey WY (1986) 'Increased volume and impaired emptying of the gallbladder in morbid obesity.' *J Nucl Med* 27: 882 (abst).
47. Shreiner DP, Sarva RP, Van Thiel D, Yingvorapant N (1986) 'Gallbladder function in diabetic patients.' *J Nucl Med* 27: 357–360.
48. Stephens CG, Scott PB (1980) 'Cholelithiasis in sickle cell anemia.' *Arch Intern Med* 149: 648–651.
49. Everson GT, Nemeth A, Kourourian S, Zogg D, Leff NB, Dixon D, Githens JH, Pretorious D (1989) 'Gallbladder function is altered in sickle hemoglobinopathy.' *Gastroenterology* 96: 1307–1316.
50. Krejs GJ, Orci L, Conlon JM, Ravazzola M, Davis GR, Raskin P, Collins SM, McCarthy DM, Baetens D, Rubenstein A, Aldor TAM, Unger RH (1979) 'Somastatinoma syndrome: biochemical, morphological and clinical features.' *N Engl J Med* 301: 285–292.
51. Fisher RS, Rock E, Levin G, Malmud L (1987) 'Effects of somatostatin on gallbladder emptying.' *Gastroenterology* 92: 885–890.

52. Beck RN, van der Vegt JH, Lichtenstein JE (1983) 'The hyperplastic cholecystosis: cholesterosis and adenomyomatosis.' *Radiology* 146: 593-- 601.
53. Swayne LS, Heitner D, Rubenstein JB, Fernandez A, Niknejad G (1987) 'Differential gallbladder contractility in fundal adenomyomatosis: demonstration by cholecystokinin cholescintigraphy.' *J Nucl Med* 28: 1771–1774.
54. Kramer EL, Rumancik WM, Harkary L, Tiu S, Banner HJ, Sander JJ (1985) 'Hepatobiliary scintigraphy of the compartmentalized gallbladder.' *AJR* 145: 1205–1206.
55. McDonald KL, Levin T (1986) 'Sonographic and scintigraphic evaluation of gallbladder duplication.' *Clin Nucl Med* 11: 692–693.
56. Masclee AAM, Jansen JBMJ, Corstens FHM, Lamers CBHW (1989) 'Reversible gallbladder dysfunction in severe pancreatic insufficiency.' *Gut* 30: 866–872.

4. Hepatobiliary imaging after gastrointestinal surgery

HEE-MYUNG PARK, HENRY N. WELLMAN and
JAMES A. MADURA

Hepatobiliary imaging with ^{99m}Tc labelled imidodiacetic acid derivatives such as ^{99m}Tc-HIDA, ^{99m}Tc-PIPIDA, ^{99m}Tc-MIBIDA, ^{99m}Tc-DISIDA has become a convenient, noninvasive and efficacious tool for the evaluation of the functional status of the hepatocytes, biliary tree, and the gallbladder. It also provides unique and important information on the biliary drainage system after varied upper gastrointestinal or biliary surgery. It plays an important role in the evaluation of postoperative complications, including postcholecystectomy syndrome. For proper interpretation of the postoperative cholescintigraphy, one must have information on the surgically altered biliary and gastrointestinal anatomy. Also, the clinical question should be well-defined.

Some of the indications for postoperative hepatobiliary imaging study are listed below (Table 1).

Enterogastric bile reflux and alkaline gastritis

Alkaline reflux gastritis is one of many postgastrectomy syndromes. The syndrome consists of burning mid epigastric pain which is unrelieved by antacids, exacerbated by meals, and worsened by assuming a recumbent position. It is associated with pyrosis, the sensation of a bitter taste in the oropharynx, usually at night [1]. With the possible exception of obvious bilious emesis, the diagnosis is not readily made on clinical grounds since there are many nonrefluxers complaining of similar symptoms.

Documentation of enterogastric reflux is relatively easy with ^{99m}Tc-IDA imaging. In an uncomplicated case, one can recognize the refluxed bile activity appearing in the usual location of the stomach, i.e. lateral to the left lobe of the liver and superior to the jejunum activity (Fig. 1). Gastric emptying is often delayed in patients with alkaline gastritis (Fig. 2). Bile reflux occurs usually between 30 min and 1 h from the time of injection, and rather abruptly. Injection of purified cholecystokinin may enhance the visualization of the reflux by dumping more bile into the duodenum. The refluxed activity

H.J. Biersack and P.H. Cox (eds), Nuclear Medicine in Gasteroenterology, 47–67

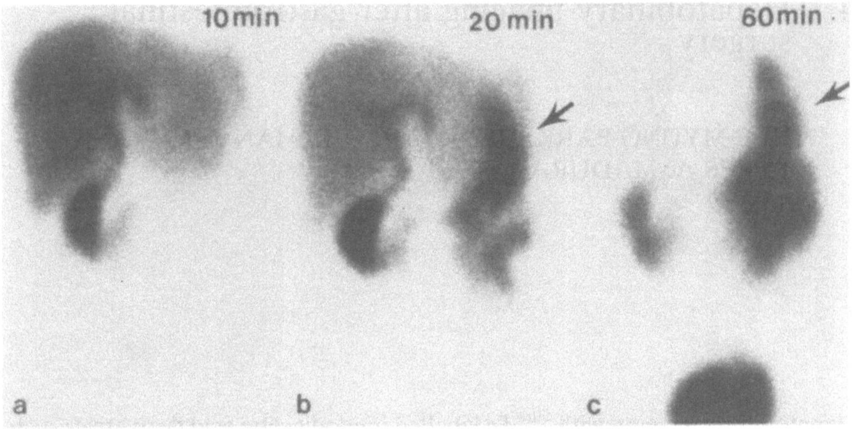

Figure 1. Enterogastric reflux and delayed gastric emptying. A 38-year-old woman who had a cholecystectomy and sphincteroplasty 5 months previously presented with persistent epigastric discomfort, nausea, vomiting and diarrhea. The ^{99m}Tc-IDA images at 10 min (1a), 20 min (1b), and 60 min (1c) show profuse bile reflux into the stomach (arrow). The stomach is well outlined by the refluxed bile activity. Up to an hour there is not much emptying of bile from the stomach (1c).

Table 1. Principle indications for postoperative ^{99m}Tc-IDA imaging

1.	Documentation and quantification of enterogastric bile reflux in suspected alkaline gastritis.
2.	Postcholecystectomy syndrome and Sphincter-of-Oddi dysfunction.
3.	Evaluation of bile flow before and after T-tube removal.
4.	Confirmation of intra- or extrahepatic biloma or bile leak suspected postoperatively or after liver injury.
5.	Documentation of biliary patency after hepatic portoenterostomy procedures for biliary atresia.
6.	Follow-up of biliary-enteric anastomosis procedures, and evaluation for biliary-cutaneous fistulae.
7.	Evaluation of liver function after partial hepatectomy.
8.	Evaluation of transplanted liver/biliary system function.
9.	Evaluation for bile ascites and bile peritonitis.
10.	Evaluation for afferent loop or efferent loop obstruction after gastroenterostomy.

may increase with time in serial ^{99m}Tc-IDA images. Occasionally jejunal activity may mimic gastric reflux. An upright image may show changing position of the jejunum due to gravity (Fig. 3). In patients with a history of complicated gastrointestinal surgery, the recognition of abnormal findings and proper interpretation of biliary imaging studies require the knowledge of the altered anatomy (Fig. 4). To confirm that a suspected activity is indeed in the stomach, one may administer a glass of water containing 0.5 to 1 mCi of ^{99m}Tc sulfur colloid or ^{99m}Tc DTPA. This will clearly outline the stomach for easy recognition. This simple technique is very helpful, particularly, in patients with surgically altered gastrointestinal anatomy (Fig. 5).

For quantification of the magnitude of enterogastric reflux, an entero-

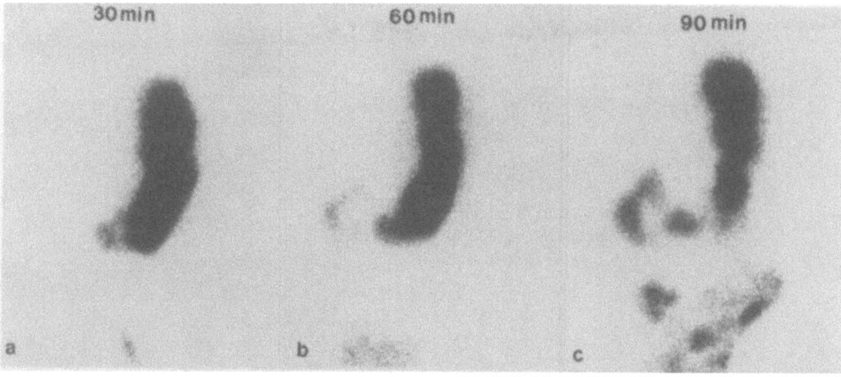

Figure 2. Enterogastric reflux and delayed gastric emptying. A 38-year-old woman who had a cholecystectomy and sphincteroplasty 5 months previously presented with persistent epigastric discomfort, nausea, vomiting and diarrhea. The 30 min (2a), 60 min (2b), and 90 min (2c) images during ^{99m}Tc-sulfur-colloid gastric emptying study show a markedly delayed emptying of labelled egg-white. Bile reflux gastritis is known to be a cause of delayed gastric emptying.

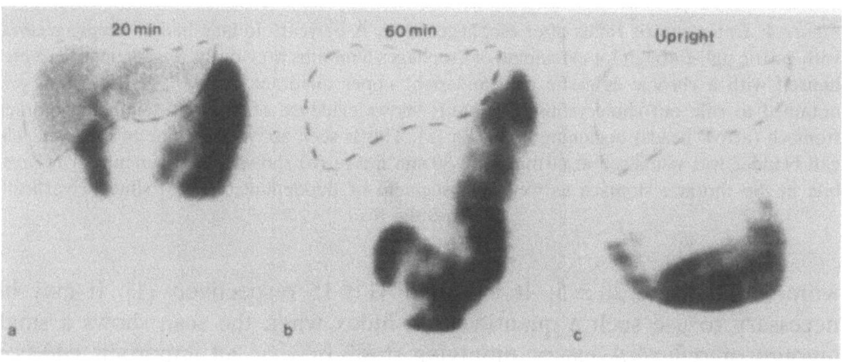

Figure 3. Jejunal activity mimicking enterogastric reflux. A 34-year-old woman who had cholecystectomy, Nissen fundoplication and sphincteroplasty several years previously presented with chronic, intermittent, epigastric pain. ^{99m}Tc-IDA study was obtained to rule out enterogastric reflux. The 20 min image (a) and 60 min image (b) shows activity in the left upper quadrant. the area mimics stomach and may possibily be misinterpreted as enterogastric reflux. The upright image (c) show the effect of gravity. Such a change indicates that it is the jejunum and not the stomach.

gastric reflux index (EGRI%) may be obtained. This requires injection of purified cholecystokinin when the ^{99m}Tc-IDA activity is considered maximal in the gallbladder. The radioactivity is monitored over the liver, gallbladder, biliary tree, small bowel, and residual stomach for 2 h. The EGRI% compares the net gain in the stomach activity after CCK with the net loss in the hepatobiliary system activity after CCK [2]. Some surgeons have found the index useful to convincingly document excessive enterogastric reflux. In one study, the mean EGRI% in control patients, nonrefluxers, and refluxers

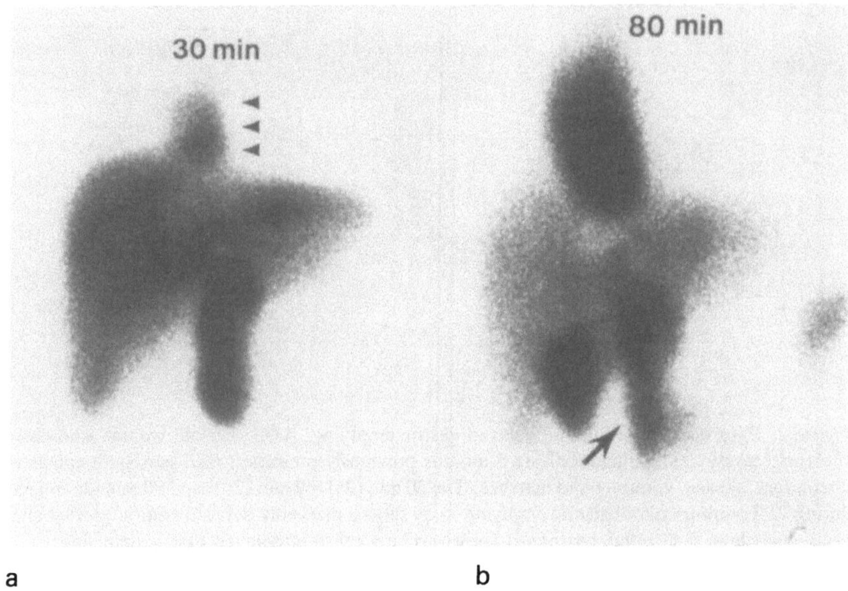

Figure 4. Enterogastric reflux after esophagectomy. A 67-year-old man had an esophagectomy with gastric pull-through for carcinoma of esophages 6 months previously. The patient presented himself with a chronic epigastic pain and right upper quadrant pain. ^{99m}Tc-IDA study was obtained to rule out cholecystitis. The study shows evidence of bile reflux into the thoracic stomach (arrow heads) beginning at 30 min (a). This is seen above the dome of the liver. The gall bladder was visualized at 60 min. The 80 min image (b) shows a large amount of refluxed bile in the thoracic stomach as well as a segment of duodenum (arrows) situated vertically behind the liver.

were found to be 20 ± 5, 16 ± 9, and 81 ± 15 respectively [1]. It may be necessary to use such a quantification index when the scan shows a small amount of reflux. A gastric emptying study may be an important addition since evidence is accumulating that gastric dysmotility may be a prominent part of alkaline gastritis.

Postcholecystectomy syndrome and sphincter-of-Oddi dysfunction

Postcholecystectomy syndrome is a poorly defined entity, characterized by recurrent abdominal pain or discomfort afflicting up to 40% of patients who have undergone cholecystectomy [3]. Forty percent of the patients with suspected recurrent biliary disease turn out to have non-biliary disorders [4]. For the narrower definition of the syndrome, the pain should be of biliary-type and associated with biochemical features of cholestasis [5]. The syndrome can result from many causes, including common duct stricture, recurrent or retained common duct stones, stone formation in a prominent cystic

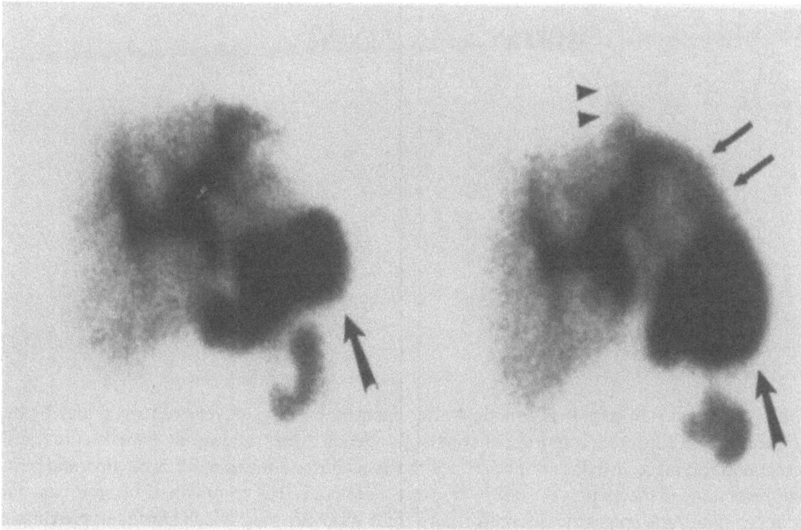

a b

Figure 5. After esophagojejunostomy. A 44-year-old woman who had a cholecystectomy, subtotal gastrectomy and Bilroth II anastomosis for peptic ulcer disease presented with persistent severe epigastric pain and melanotic stool. Endoscopy revealed a severe gastritis and esophagitis. Subsequently a total gastrectomy and esophagojejunostomy was performed. ^{99m}Tc-IDA study was obtained to see if there was enteroesophageal bile reflux. The 60 min image (a) shows the bile in the jejunal pouch (arrow) but no suggestion of reflux into the esophagus. To visualize the location of the esophagus and the jejunal pouch, a glass of water containing 1 mCi of ^{99m}Tc-IDA was given to the patient. The image shows the locations of the esophagus (arrow heads), the jejunal conduit (short arrows), and the jejunal pouch (arrow) (b).

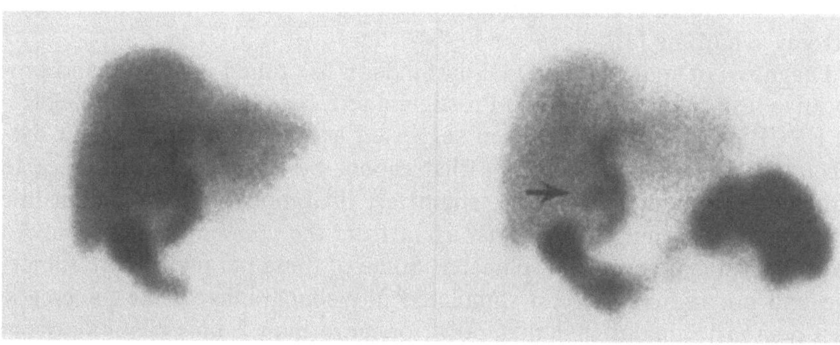

a b

Figure 6. Cystic duct remnant. A 38-year-old woman who had a cholecystectomy previously presented with intermittent epigastric pain and vomiting. ^{99m}Tc-IDA study was obtained to rule out any biliary problem. The 20 min (a) and 45 min image (b) shows a small cystic duct remnant (arrow). Otherwise the liver and biliary system were normal. Rarely a stone can be formed in the cystic duct remnant and cause postcholecystectomy syndrome.

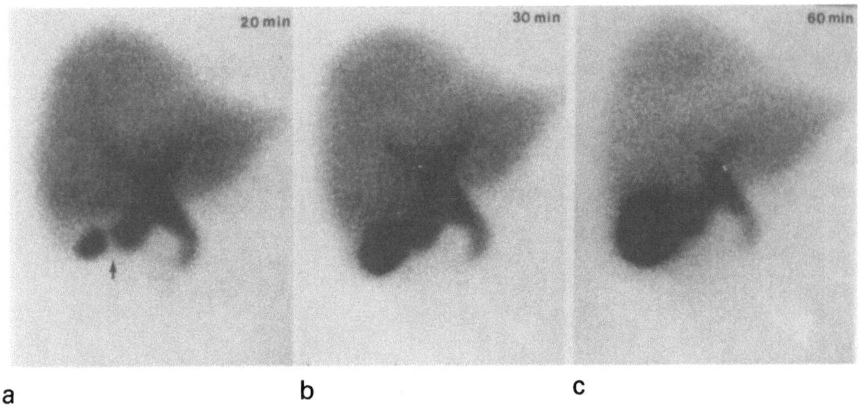

Figure 7. Sphincter of Oddi dysfunction after laparotomy. A 68-year-old man had ligation and resection of ruptured inferior pancreaticoduodenal artery aneurysm 6 weeks previously. Pancreaticoduodenectomy was not needed for the procedure. An extensive dissection and partial wedge resection of the pancreas, however, were necessary. He complained of anorexia, mild right upper quadrant/epigastric discomfort and right shoulder pain which started approximately 2 weeks after the surgery. Abdominal CT showed a low density structure consistent with enlarged gallbladder or an abscess. This was shown by ^{99m}Tc-IDA study to be a large gallbladder. There was prompt filling of the gallbladder, which, incidentally, appeared septated (a, b). The common bile duct is prominent with stasis of bile (c). Persistence of bile stasis for many hours would have indicated obstruction. In this patient, bile activity emptied into the small bowel at 4 h image suggesting sphincter dysfunction but no obstruction. This was confirmed by subsequent ERCP.

duct remnant (Fig. 6), or stenosis of the sphincter-of-Oddi [4]. Also included as causes are pancreatitis and neoplasm [5]. Ampullary stenosis (stenosis of the sphincter-of-Oddi, papillary stenosis, papilla of Vater stenosis), with or without associated choledocholithiasis, is a major cause of the postcholecystectomy syndrome [4].

Diagnosis of sphincter-of-Oddi dysfunction has relied on ERCP and provocative tests like the morphine-Prostigmin test. Several findings on diagnostic ERCP examination have been suggested as indicators of sphincter dysfunction. These include a papilla that cannot be cannulated, resistance to catheter movement through the sphincter, dilatation of the common bile duct, delayed drainage of contrast agent from the common bile duct and a pressure gradient across the sphincter. Some of these parameters are subjective and not useful. Secretin stimulated ultrasonography of the pancreatic duct demonstrating greater than 50%, or more than 2 mm, dilatation from baseline along with abdominal pain has several proponents as a reliable indicator of stricture or spasm at the sphincter-of-Oddi. Dilatation of the common bile duct, the Morphine-Prostigmin test and liver-function tests were shown to have negligible value in predicting whether a patient would benefit from sphincterotomy [6]. Manometric ERCP appears to be the most reliable indicator but requires an invasive catheterization procedure.

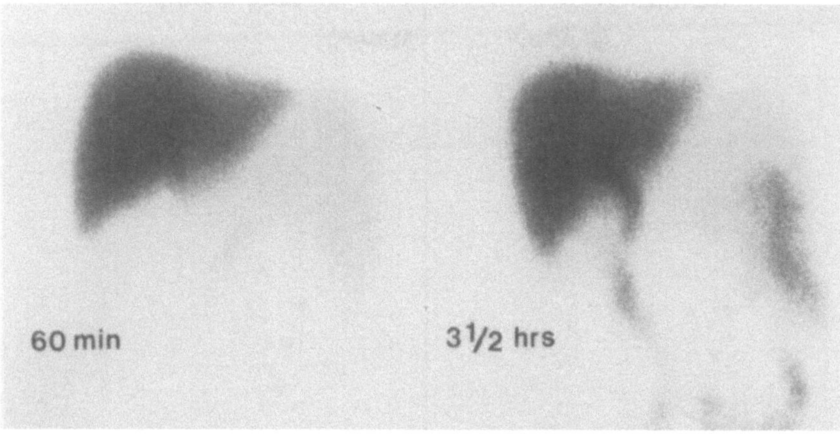

a b

Figure 8. Postcholecystectomy syndrome due to Sphincter-of-Oddi dysfunction. A 46-year-old woman had a cholecystectomy two years previously and has been complaining of dull aching right upper quadrant pain which is not relieved by antiacid, also diarrhea and fatigue for 6 months. ^{99m}Tc-IDA study was performed to rule out sphincter dysfunction. The study shows good hepatic uptake but very slow excretion of bile into the gut. The 60 min image (a) shows no activity in the small bowel. The lack of activity in the common bile duct is probably due to its high pressure. Delayed image at $3\frac{1}{2}$ h (b) shows activity in the gut indicating sphincter dysfunction but no obstruction.

Cholescintigraphy has had significant impact on the evaluation of the postcholecystectomy syndrome. The technique allows physiologic assessment of duct drainage and usually correlates well with the washout of contrast material from the biliary tract as observed on endoscopic retrograde cholangiopancreatography [7]. Dysfunction or obstruction of the sphincter can readily be identified by biliary scintigraphy (Figs 7, 8). The differentiation of functional obstruction (e.g., spasm) from pathological obstruction, however, usually relies on ERCP. Ultrasound or CT study cannot distinguish permanent nonobstructive ductal dilatations from obstruction, whereas, biliary scintigraphy can easily identify obstruction. For these reasons, some recommend biliary scintigraphy as the initial screening study to preselect patients for ERCP [7].

Efforts have been made to find the most reliable quantitative parameter in ^{99m}Tc-IDA scintigraphy. Some found that the optimal variable was the hepatic clearance at 45 or 60 min after injection of the radiotracer. The clearance rates from the right lobe of the liver in true negative cases and in true positive cases were 70 ± 9% and 51 ± 20% respectively. The sensitivity and specificity of this parameter were 67% and 85% respectively [8]. Others found that the most reliable criterion was the time at which maximal bile duct activity occurred. The times-to-peak (bile duct) in unobstructed group and in obstructed group were 28 ± 10.1 and 44.1 ± 14.3 min respectively.

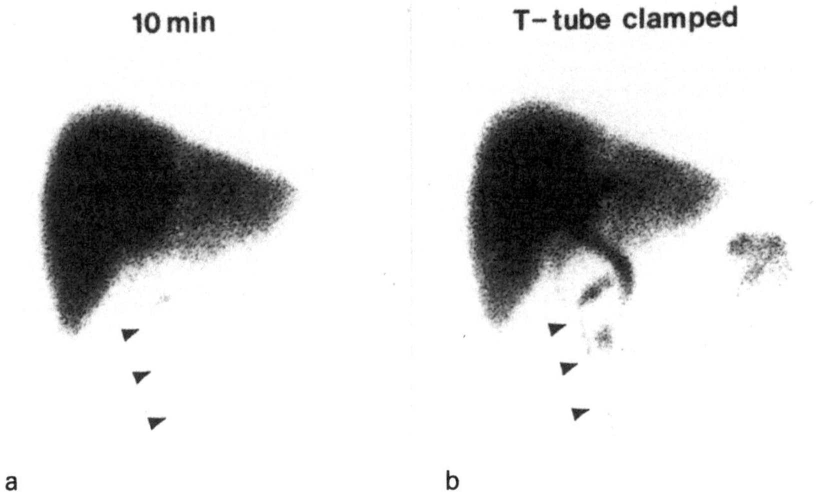

Figure 9. Bile flow after clamping T-tube. A 58-year-old man had a successful liver transplantation which was done due to liver failure secondary to α-antitrypsin deficiency cirrhosis. ^{99m}Tc-IDA study was obtained to see the patency of the biliary tree when the T-tube is clamped. The 10 min image (a) shows a rapid uptake in the transplanted liver with a rapid excretion of bile into the T-tube (arrow heads). Another image obtained after the T-tube was clamped (b) shows a normal biliary flow into the common bile duct and into the small bowel confirming biliary patency.

Using a cutoff level of 29 min or more as a criterion, the sensitivity and specificity were 93% and 64% respectively and an overall accuracy of 80% in the prediction of obstruction [9].

Bile flow before and after T-tube removal

T-tubes become dislodged for a variety of reasons including accidental pulling. The most common reason for T-tube dislodgement is a slow, indolent migration into the sinus tract caused by breathing and normal movement of the distal limbs [4]. When a T-tube is completely removed inadvertently and the injection of contrast into the catheter tract does not opacify the biliary tree, ^{99m}Tc-IDA scanning becomes a very useful tool. The study will provide necessary information as to the patency of the biliary tree, and will demonstrate bile leakage into the T-tube tract if present. It is also a very useful procedure before a planned removal of T-tube. The tube should be clamped, and followed by a ^{99m}Tc-IDA biliary scintigraphy to see if the biliary tree is patent (Fig. 9).

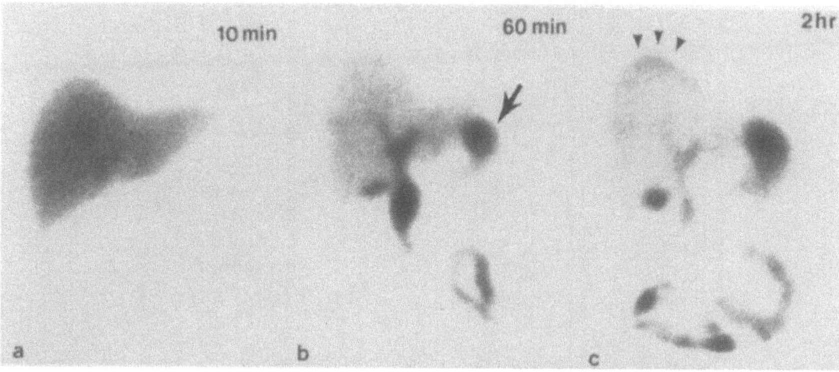

Figure 10. Bile leak after transhepatic stent placement. A 45-year-old woman with cervical carcinoma had a common bile duct obstruction secondary to a peripancreatic tumor mass. An internally draining stent was placed in the common bile duct following a transhepatic cholangiogram. She developed abdominal pain and fever. ^{99m}Tc-IDA study was obtained to rule out bile leak. The 10 min image (a) shows a good hepatic uptake. The 60 min image (b) shows activity in the gallbladder, large enterogastric reflux (arrow heads). A delayed image obtained at 2 h shows the bile over the dome of the liver (arrow heads). The visualization of the bile under the diaphragm is enhanced by the clearing of the activity from the liver (c).

Biloma and bile leakage

Injuries to the biliary system during cholecystectomy are relatively infrequent. The incidence is between 0.1 and 0.7%, averaging approximately one for every 500 cholecystectomies [10]. The two most common injuries are biliary stenosis and leakage from either the cystic duct or an accessory bile duct in the gallbladder bed which drains anomalously into the gallbladder. The leakage is either localized within or around the liver, or diffuse throughout the peritoneal cavity. The pathogenesis of bilomas is due to a bile leak, which is 'walled off,' resulting in the formation of a bile cyst. These are typically encapsulated and located in the right upper quadrant near the liver and biliary tree [11]. Infection and rupture of biloma into the peritoneal cavity need aggressive surgical management [12].

Computed tomography and ultrasonography are excellent, sensitive methods of detecting abnormal fluid collections in the abdomen. Biliary scintigraphy is a most specific, physiologic, noninvasive, and effective means of confirming biloma. Early detection can significantly reduce morbidity and mortality [13]. The usual scintigraphic findings in bile leakage are normal liver function and demonstration of radioactivity outside the biliary tract. Different patterns can be seen depending on the type and extent of the bile leakage [14]. The radioactive bile can spread surrounding the liver, often accumulating under the right hemidiaphragm (Fig. 10); it can be localized in the gallbladder fossa (Fig. 11); or it can distribute throughout the peritoneal cavity (Fig. 21). If the biloma is sealed off completely and not communicating with the bile, it may appear as a photopenic defect.

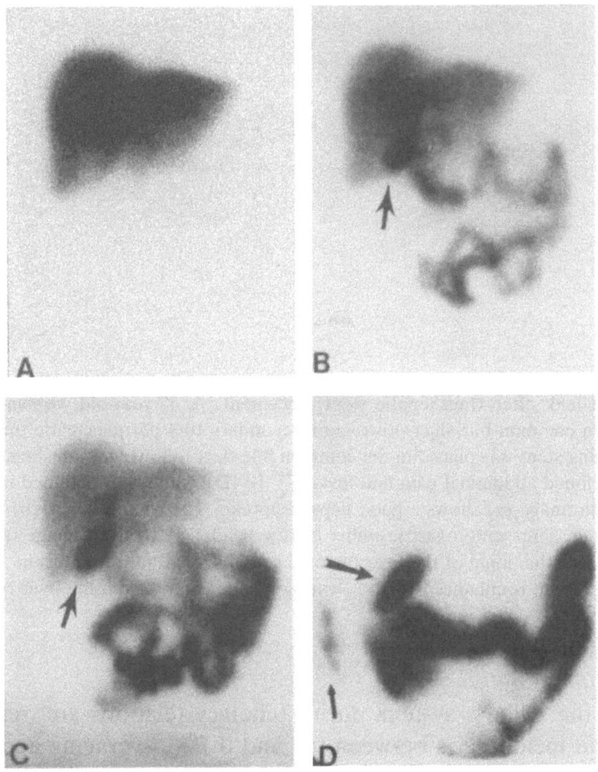

Figure 11. Bile leak after cholecystectomy. A 58-year-old woman had a recent cholecystectomy and a continuous drainage from a catheter placed in the region of the gallbladder. She complained of upper abdominal pain, nausea and vomiting. Ultrasound examination showed a fluid collection in the right upper quadrant. ^{99m}Tc-IDA images obtained at 5, 20, 40 min and 6 h (a, b, c, d) show normal liver function, prompt bile excretion and a persistent abnormal collection of activity in the gallbladder fossa indicating localized bile leak (arrow). The six hour image also shows some activity in the external collection bag (small arrow). (Reproduced with permission of the publisher from Siddiqui A. (1986): 'Different patterns for bile leakage following cholecystectomy demonstrated by hepatobiliary imaging.' *Clin Nucl Med* 11: 751–753.)

After hepatic portoenterostomy

Kasai and Suzuki introduced an operation for 'noncorrectable' biliary atresia in 1959. The basic concept of the surgical approach is based on the fact that there are minute biliary channels in the fibrous tissue at the porta hepatis. The operation consists of resection of this fibrous tissue and anastomosis of a Roux-en-Y loop of the jejunum to the resected area. Without surgical drainage of the porta hepatis, the 4-year survival rate is 2%; whereas, with surgery a 42% 4-year survival rate may be achieved [15]. A primary determinant in survival appears to be adequate bilirubin secretion [15]. Hepatobiliary imaging with ^{99m}Tc-IDA compounds has been shown to be useful in deter-

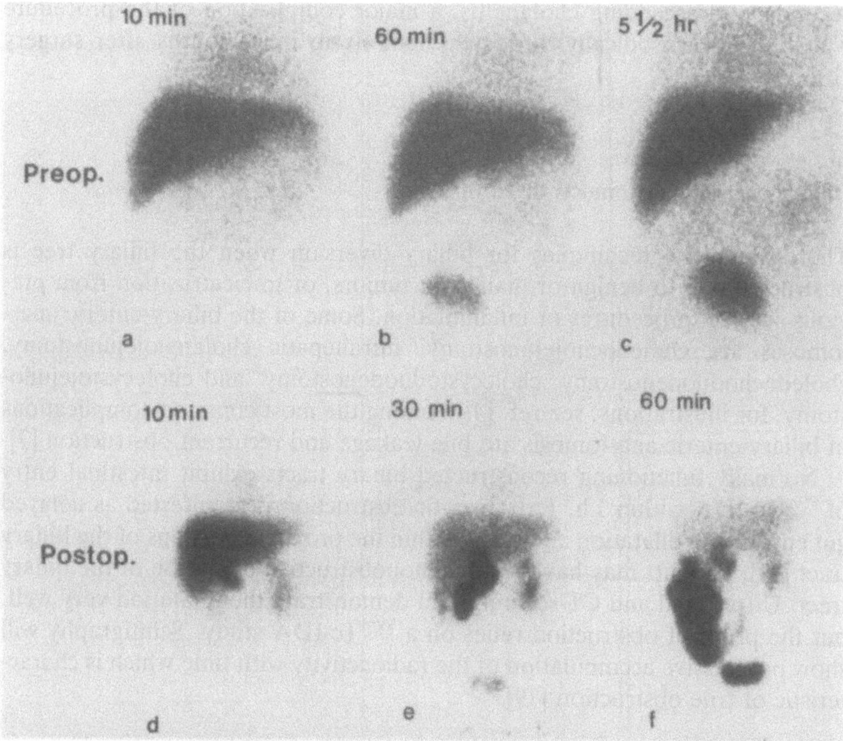

Figure 12. Before and after a portoenterostomy for biliary atresia. A 5-week-old boy with hyperbilirubinemia and increased liver function tests had a ^{99m}Tc-IDA scan to rule out biliary atresia. The 10 min (a), 60 min (b), and $5\frac{1}{2}$ h (c) images show good hepatic uptake but no excretion of bile into intestine, consistent with biliary atresia. After Kasai portoenterostomy, ^{99m}Tc-IDA study was repeated. The images at 10 min (d), 30 min (e), and 60 min (f) after injection show a good hepatic uptake with a rapid excretion of the bile into the jejunal loop as early as in 10 min. By 60 min most of the activity is cleared out of the liver into the intestine indicating a widely open anastomosis. There is no abnormal accumulation of bile to suggest leakage.

mining the status of bile flow before and after surgery. There have been several reports on methods of quantification of hepatobiliary function. It has also been shown that an estimated bile bilirubin clearance index derived from the inscribed hepatic absorption and elimination curves are a reliable indicator of bilirubin clearance [16].

^{99m}Tc-DISIDA hepatobiliary imaging is an efficient method of detecting the integrity and complications in patients undergoing hepatic portoenterostomy [17]. A successful operation is indicated by a prompt excretion of bile into the jejunal loop and beyond (Fig. 12). Retention of bile in the intrahepatic bile ducts or in the liver parenchyma without excretion into the jejunal conduit indicates nonpatency of the portoenterostomy. Nonpatency may be due to biliary canaliculi being too small, postoperative stricture, or secondary to ascending cholangitis, a major complication of the procedure which occurs episodically during the first six to nine months after surgery [18].

Biliary-enteric anastomosis in adult

There are many techniques for biliary diversion when the biliary tree is obstructed due to benign or malignant tumors, or to cicatrization from previous surgical procedures or inflammation. Some of the biliary-enteric anastomoses are choledochojejunostomy, intrahepatic cholangiojejunostomy, choledochoduodenostomy, cholecystoduodenostomy, and cholecystojejunostomy, for illustrations, see ref. [4]. Among the most common complications of biliary-enteric anastomosis are bile-leakage and recurrent obstruction [7].

Normally functioning reconstructed biliary tracts exhibit intestinal entry of ^{99m}Tc-IDA within 1 h. Extrahepatic obstruction is manifested as delayed gut entry, with dilatation and stasis within the proximal portions of the biliary tract [19]. Patients may have residual nonobstructive dilatation of the biliary tract. Ultrasound and CT scanning can demonstrate the dilatation very well, but the proof of obstruction relies on a ^{99m}Tc-IDA study. Scintigraphy will show progressive accumulation of the radioactivity with time which is characteristic of true obstruction [19].

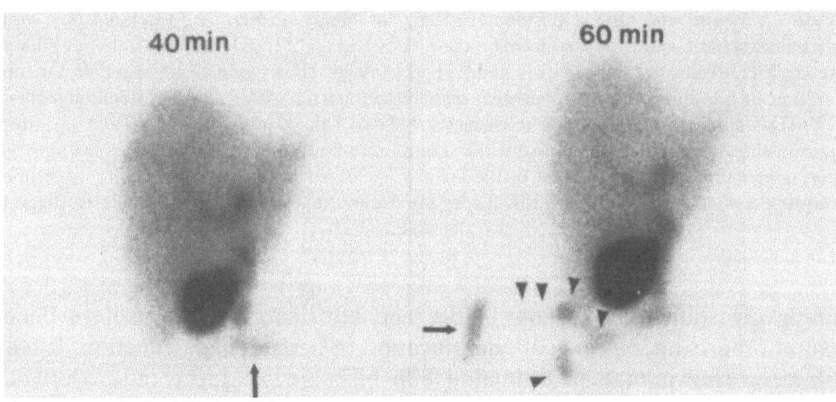

Figure 13. Bile leak through duodenal fistula. A 70-year-old woman with carcinoma of the colon had right colectomy with resection of second portion of duodenum. She developed complications including duodenal fistula and pelvic abscess. The fistula was repaired but there was a persistent drainage from the abdominal wound. ^{99m}Tc-IDA scan was obtained to evaluate the flow of bile. There was prompt visualization of the gallbladder, and the 40 min image (a) shows bile activity in the duodenum (arrow). The 60 min image (b) shows abnormal activities (arrow heads) extending laterally to the outside of the abdominal cavity where a drainage bag was located (arrow). The study suggested that the leakage was probably from the duodenum. It was confirmed at surgery.

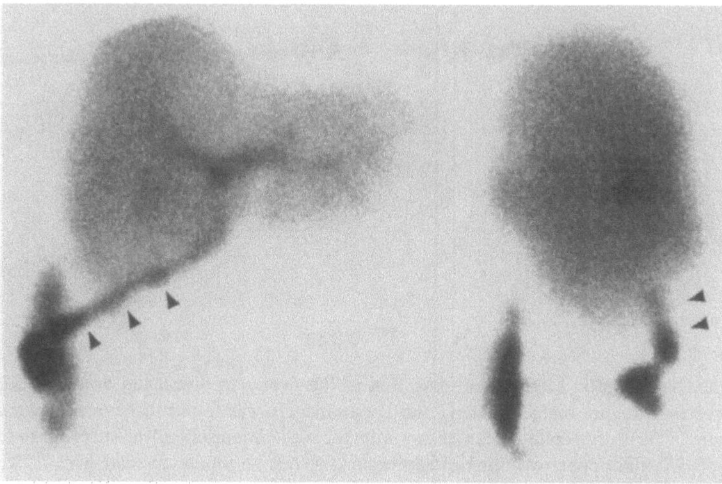

Figure 14. Biliary cutaneous fistula following Whipple procedure. A 17-year-old boy was hit by a car while riding a snowmobile. He received numerous injuries including multiple abdominal organ injury requiring Whipple procedure. He developed postoperative enterocutaneous fistula which was treated with central hyperalimentation and drainage. ^{99m}Tc-IDA study was obtained to see if there was any drainage of bile into the small bowel. The 10 min image (a) shows good hepatic uptake and biliary drainage through the enterocutaneous fistula (arrow heads) into the ostomy bag. A lateral view (b) shows the relative anterior location of the fistula tract (arrow heads) and the more posterior location of the fluid level in the ostomy bag. Images up to 60 min showed no bile activity in the intestine.

^{99m}Tc-IDA hepatobiliary imaging has been shown to be a very effective modality to confirm active intradominal biliary leakage (Fig. 13), as well as bile leak through biliary-cutaneous fistula (Fig. 14). Biloma appears as a cold (photopenic) defect initially, and with time it becomes 'hot' due to mixing with radioactive new bile. Occasionally, the tracer may initially accumulate in the gall-bladder fossa mimicking normal visualization of the gallbladder [20] (Fig. 11).

Partial hepatectomy

Liver scanning is useful in evaluating hepatic regeneration after resection. After partial hepatectomy, the liver regenerates very rapidly in children, reaching the normal volume, but not shape, in one to three months [21]. When a hepatic lobe is removed portal flow to the liver remnant will be increased immediately. Increase in its bulk due to vascular engorgement and edema accounts for the earliest change, and the later change is true regenerative hyperplasia which will contribute to the restoration of normal liver function [22]. Histologic evidence of regeneration including increased mitotic

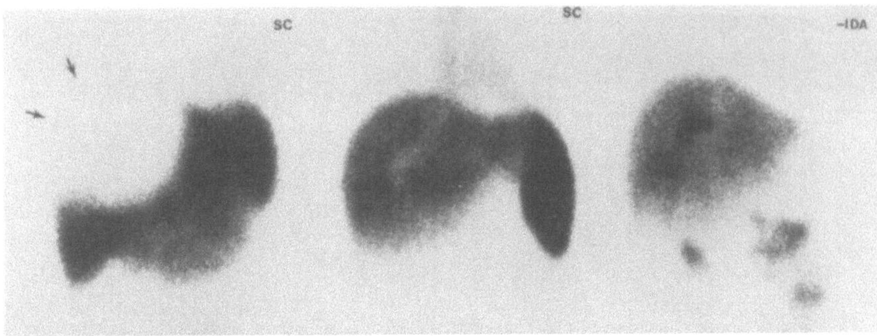

a b 17 days c

Figure 15. Regeneration after partial resection of the liver. An 8-year-old boy who presented with increasing fatigue, malaise, fever, and hepatomegaly was found to have a hepatocellular carcinoma. ^{99m}Tc-sulfur colloid scan shows a large, well-circumscribed photopenic defect (arrows) replacing the upper two-third of the right lobe (a). A follow-up scan with ^{99m}Tc-sulfur colloid (b) was obtained 17 days after a right hepatic lobectomy and cholecystectommy. The liver already resumed a normal shape and size. Note the interim enlargement of the spleen presumably due to a relative increase in portal venous pressure after right lobectomy.

^{99m}Tc-IDA scan obtained 25 days after right lobectomy show the liver is now normal in size, shape, and in biliary function (c). The bile activity in the small bowel is seen as early as 15 min after the injection. The gallbladder is surgically absent.

activity has been observed within three days [23]. ^{99m}Tc sulfur colloid has, traditionally, been used to document regeneration of the liver (Fig. 15). ^{99m}Tc-IDA is an excellent agent to image the liver after partial hepatectomy. Not only will the size of the regenerating liver be shown but its functional status, including biliary flow (Fig. 15c). ^{99m}Tc-IDA is the agent of choice to rule out bile leakage when there is perihepatic fluid collection (Fig. 16). It is a reliable, noninvasive postoperative study to rule out iatrogenic bile duct injury should the patient develop hyperbilirubinemia.

Transplanted liver

Since the National Institutes of Health consensus meeting in 1983 concluded that liver transplantation should be considered a therapeutic modality for end-stage liver disease, this procedure has become a reasonable option for treatment of many patients with the liver disease. Patients with primary sclerosing cholangitis, chronic active hepatitis, and primary biliary cirrhosis have the best l-year survival rates (90%, 84%, and 85%, respectively), whereas the corresponding rate for those with acute hepatitis was 63% [24]. Rejection, infection, biliary tract complications and hepatic artery thrombosis, in the order of frequency, are the common complications of liver transplantation.

Hepatobiliary scintigraphy has a proven role in assessing the patency of

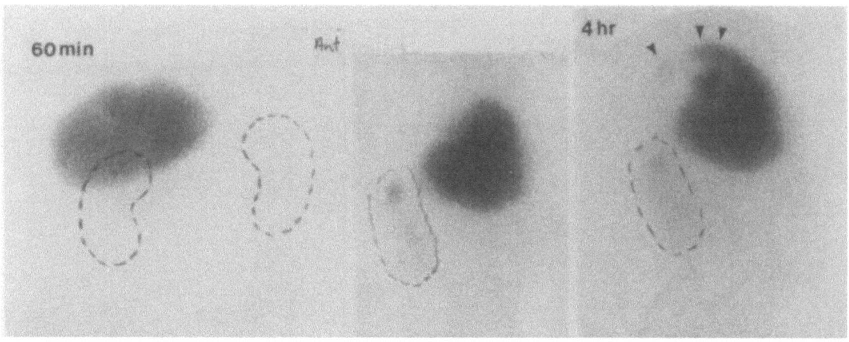

a b Lat. c Lat.

Figure 16. Obstruction and leakage of bile flow after multiple segmentectomy of liver. A 67-year-old man who had a left colectomy for colon carcinoma presented with new liver metastases. After a hepatic trisegmentectomy (right anterior, right posterior and left medial segments) and cholecystectomy were performed, the total bilirubin rose from 4.6 to 14.2 mg in 4 days. ^{99m}Tc-IDA scan was obtained to evaluate the patency of the biliary ductal system. Up to 60 min into the study (a), there is no identifiable bile duct, or evidence of bile flow into small bowel. The kidneys (dotted outlines) were visualized on a posterior view. The liver appears somewhat flattened but already assuming a near normal shape. A lateral view (b) shows that the liver is flattened superior-posteriorly. Four-hour delayed lateral view (c) shows bile activity appearing on the superior-posterior aspect of the liver and under the diaphragm (arrow heads) indicating bile leakage. There was no activity in the intestine.

transplanted anastomosed hepatic bile ducts (Fig. 17) and bile leakage (Fig. 18). The ^{99m}Tc-IDA scan has been helpful in differentiation of rejection crisis (Fig. 19) and extrahepatic biliary obstruction [25]. Serial scintigraphic studies allow monitoring of any changes in hepatic blood flow or function (Fig. 20). ^{99m}Tc-IDA studies obtained during episodes of rejection clearly show a slowing of extraction and clearance of IDA, but this observation is not specific for rejection in as much as other causes of depression of hepatic function such as viral hepatitis result in similar changes [26]. Various quantification parameters are being evaluated to assess the changes in transplant hepatic function [27, 28].

Bile peritonitis and ascites

Chronic bile peritonitis and bile ascites can occur as a complication of a percutaneous biliary drainage procedure or cholecystectomy [11, 14, 29]. ^{99m}Tc-IDA scintigraphy can clearly demonstrate the radioactive bile mixing with the ascitic fluid. For this purpose, a delayed image must be obtained several hours after the injection (Fig. 21).

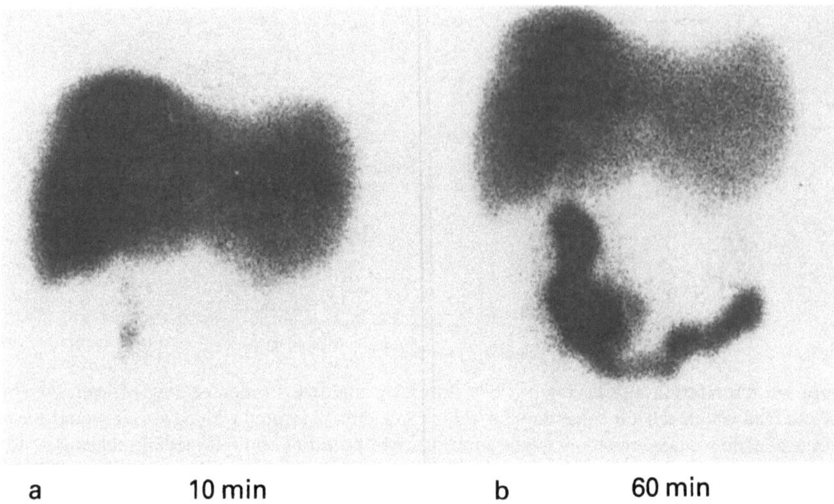

Figure 17. Patent biliary system after transplantation. A 40-year-old woman received a cadaveric liver transplantation for her end-stage-liver disease secondary to sclerosing cholangitis. ^{99m}Tc-IDA study was obtained to evaluate the biliary patency. The T-tube was clamped before injection. The 10 min image (a) and 60 min image (b) show excellent hepatic uptake and prompt excretion of bile into the small bowel without bile leakage indicating successful transplantation.

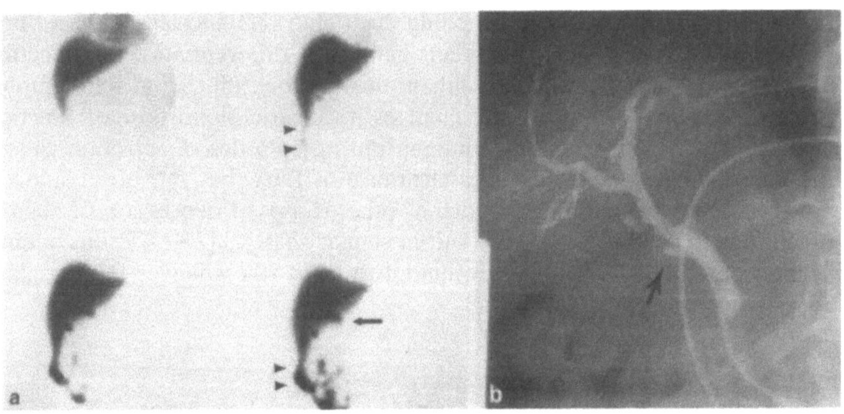

Figure 18. Bile leak after liver transplantation. ^{99m}Tc-IDA scan after removal of T-tube in a 45-year-old man who had a liver transplantation. He complained of increasing pain in the right upper and right lower quadrant of abdomen after the T-tube was removed. The 10 min image (a) shows a linear activity below the tip of the liver extending downward along the 'gutter' suggesting bile leakage (arrow heads). This activity increases with time and seeps between the loops of bowel. There is little activity draining into the duodenum from the common bile duct (arrow). The leakage was confirmed by injecting contrast material through the nasobiliary tube (b). The leakage is from the site of previous T-tube insertion (arrow), and is seen below the donor cystic duct remnant.

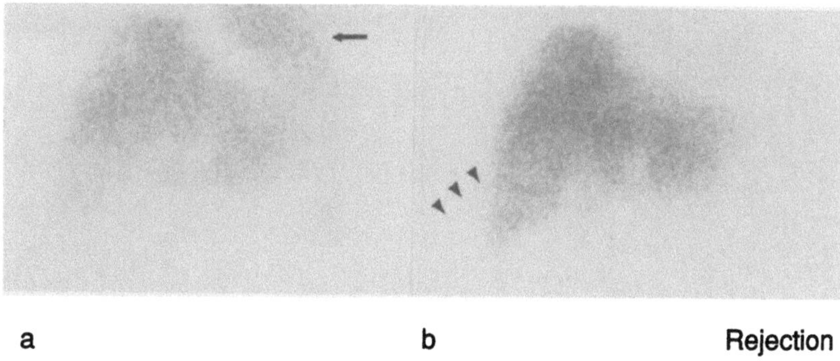

Figure 19. Transplant rejection. A 56-year-old woman had a liver transplantation for liver failure due to chronic active hepatitis. On sixth postoperative day, she became increasingly somnolent with fever. ^{99m}Tc-IDA scan was obtained to evaluate possible bile leak or transplant rejection. The 10 min image (a) shows very poor hepatic uptake with persistent activity in the cardiac chamber (arrow). Even after 60 min (b) there is not much liver uptake. Biliary trees are not visualized and there is very little excretion of bile in the T-tube (arrow heads). These findings are consistent with rejection phenomenon.

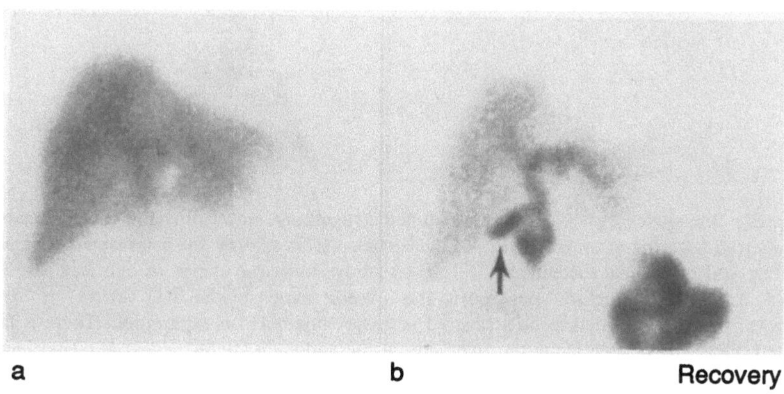

Figure 20. Recovery from rejection episode. The same patient as in Fig. 19, after recovery from a rejection episode. The 10 min (a) and 60 min (b) images show markly improved hepatic uptake and clearance with prompt visualization of the biliary trees. The proximal portion of the T-tube is also visible (arrow).

Afferent loop or efferent loop obstruction

Patients with partial gastric resection with gastrojejunostomy (Bilroth II anastomosis) may rarely suffer from 'afferent loop syndrome.' Most common complaints are abdominal bloating and pain 20 min to 1 h after eating, frequently followed by nausea and vomiting. Characteristically, the bloating

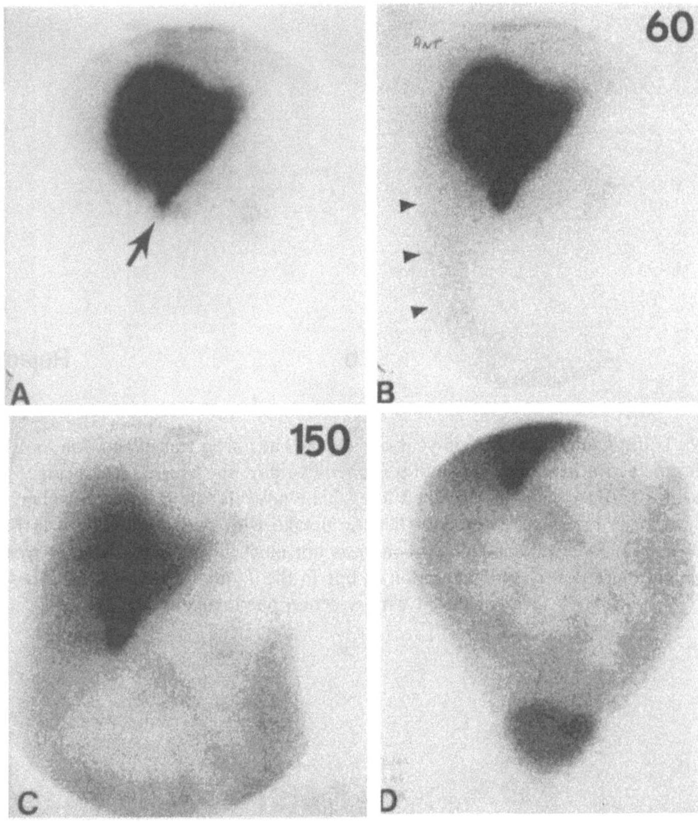

Figure 21. Bile ascites. A 66-year-old woman had a vagotomy, antrectomy and cholecystectomy for penetrating gastric ulcer and gallstones. One week after discharge, she complained of nausea, vomiting and abdominal distention. ^{99m}Tc-IDA study showed evidence of bile leak and bile ascites. A tube-like structure appeared at the inferior margin of the liver (arrow) at 10 min image (a), which was found to be a severed accessory bile duct on exploration. There is faint (diluted bile) activity seen along the right side of the abdomen (arrow heads) at 60 min image (b). At 150 min (c, d), the bile is seen distributed throughout the peritoneum mixed with ascitic fluid (bile ascites). (Reproduced with permission of the publisher from Siddiqui A. (1986) 'Different patterns for bile leakage following cholecystectomy demonstrated by hepatobiliary imaging.' *Clin Nucl Med* 11: 751–753.)

and abdominal discomfort are relieved by vomiting. This type of afferent loop syndrome is considered to be caused by distention of a partially obstructed afferent loop by bile and pancreatic secretions which are stimulated by eating [30].

In patients with uncomplicated gastroenterostomy, the afferent and efferent loops are visualized with ^{99m}Tc-IDA within 1 h. Most patients with Billroth II and Whipple's procedures will portray gastric reflux, whereas it is considerably less common in gastroenterostomy plus a Roux-en-Y procedure [19]. The washout from the afferent loop is normally completed within

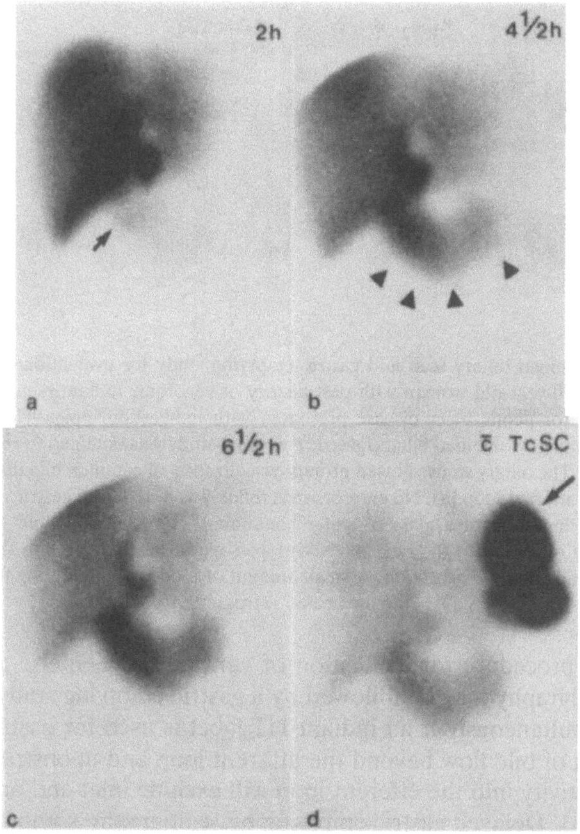

Figure 22. Afferent loop syndrome. A 58-year-old man had partial gastric resection and Bilroth II for peptic ulcer disease one year previously. He also had cholecystectomy. The patient presented with postprandial abdominal bloating and pain, which were usually followed by nausea and vomiting. ^{99m}Tc-IDA study was obtained to see the status of bile flow through the afferent loop. The study showed excretion of bile into the afferent loop of the bowel (a, arrow; b, arrow heads). The activity, however, remained stagnant in the loop without moving further down into the efferent loop up to $6\frac{1}{2}$ h of observation (c). A glass of water with 1.0 mCi of ^{99m}Tc-sulfur colloid was given to identify the location of the stomach (d, arrow). These findings suggested afferent-loop obstruction, which was confirmed by subsequent endoscopy.

2 h. Using these temporal parameters, it is possible to disclose and distinguish afferent loop obstruction, efferent loop obstruction, and inlet-outlet obstruction about the gastroenteric anastomosis [19]. Afferent loop obstruction can be easily demonstrated by ^{99m}Tc-IDA study (Fig. 22). Administration of a glass of water containing ^{99m}Tc-sulfur colloid or DTPA at the end of the study will help locate and outline the gastric remnant (Fig. 22d). A biliary study alone may be inadequate to evaluate the outlet obstruction if there is not enough bile flow into the small bowel to visualize the efferent loop. A combined biliary/gastric emptying study is quite helpful and it may well be

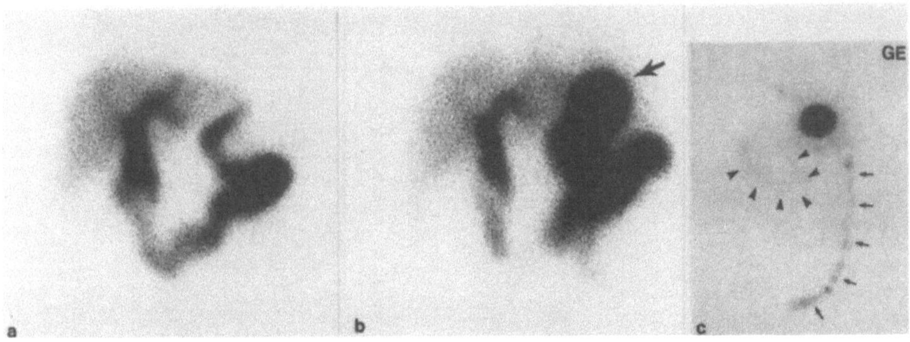

Figure 23. Combined biliary scan and gastric emptying study for evaluation of inlet or outlet obstruction. A 70-year-old woman with past history of vagotomy and antrectomy with Billroth II anastomosis for peptic ulcer disease presented with indigestion, epigastric discomfort and occasional vomiting. Combined biliary/gastric emptying study was obtained to rule out inlet/outlet obstruction. The biliary study showed prompt visualization of common bile duct and emptying of bile into the afferent loop (a). No enterogastric reflux was noted. The gastric remnant (arrow) is shown after ingestion of a glass of water containing ^{99m}Tc-sulfur colloid (b). The gastric emptying study (c) obtained the next day shows passage of labeled egg into the efferent loop (arrows) excluding outlet obstruction. A small amount of food is seen entering the afferent loop (arrow heads) retrogradely.

the routine procedure for evaluation of various gastroenteric anastomoses. Biliary scintigraphy may be followed by a gastric emptying study or both can be done simultaneously if an indium-111 label is used for gastric emptying. Observation of bile flow beyond the afferent loop and unobstructed passage of gastric activity into the efferent loop will exclude inlet and outlet obstruction (Fig. 23). Delayed gastric emptying by scintigraphy without obstruction seen by endoscopy should indicate abnormal gastric motility as the cause.

References

1. Ritchie WP Jr (1986), 'Alkaline reflux gastritis: late results on a controlled trial of diagnosis and treatment.' *Ann Surg* 203: 537–544.
2. Tolin RD, Malmud LS, Stelzer F et al. (1979) 'Enterogastric reflux in normal subjects and patients with Billroth II gastroenterostomy.' *Gastroenterol* 77: 1027–1033.
3. Bodvall B (1973) 'The post-cholecystectomy syndromes. In: *Clinics in Gastroenterology*. Vol. 2. pp. 103–126. London: WB Saunders.
4. Zeman RK, Burrell MI (1987) 'The postoperative biliary tract.' In: *Gallbladder and Bile Duct Imaging*. pp. 593–675. New York: Churchill Livingstone.
5. Shaffer EA, Hershfield NB, Logan K, Kloiber R (1986) 'Cholescintigraphic detection of functional obstruction of the sphincter of Oddi.' *Gastroenterol* 90: 728–733.
6. Greenen JE, Hogan WJ, Dodds WJ et al. (1989) 'The efficacy of endoscopic sphincterotomy after cholecystectomy in patients with sphincter-of-Oddi dysfunction.' *N Engl J Med* 320: 82–87.
7. Zeman RK (1988) 'Correlation of nuclear techniques with other hepatobiliary imaging modalities.' In: Gottschalk A, Hoffer PB, Potchen EJ (eds), *Diagnostic Nuclear Medicine*. pp. 621–630. Baltimore: Williams & Wilkins.

8. Darweesh, RM, Dodds WJ, Hogan WJ et al. (1988) 'Efficacy of quantitative hepatobiliary scintigraphy and fatty-meal sonography for evaluating patients with suspected partial common duct obstruction.' *Gastroenterol* 94: 779–786.
9. Kloiber R, AuCoin R, Hershfield NB et al. (1988) 'Biliary obstruction after cholecystectomy: Diagnosis with quantitative cholescintigraphy.' *Radiol* 169: 643–647.
10. Seror J, Schmitt JC, Patras CL, et al. (1978) 'Operative injuries to the bile ducts: report on 12 cases.' *Int Surg* 63: 108.
11. Makhija M, Schultz S, McManus A (1985) 'Scintigraphy of bile leakage following cholecystectomy.' *Clin Nucl Med* 10: 848–850.
12. Alberti-Flor J (1984) 'Biloma: an unusual post cholecystectomy complication.' *Contemporary Surg* 24: 108.
13. Weissman H, Chun K, Frank M, et al. (1979) 'Demonstration of traumatic bile leakage with cholescintigraphy and ultrasonography.' *AJR* 133: 843.
14. Siddiqui A, Ellis J, Madura J (1986) 'Different patterns for bile leakage following cholecystectomy demonstrated by hepatobiliary imaging.' *Clin Nucl Med* 11: 751–753.
15. Hitch DC, Shikes RH, Lilly Jr (1979) 'Determinants of survival after Kasai's operation for biliary atresia using actuarial analysis.' *J Pediatr Surg* 14: 310–314.
16. Hitch DC, Leonard JC, Manion CV, et al. (1981) 'Assessment of hepatic function after hepato-porto-enterostomy for biliary atresia using Tc-Diethyl-IA.' *J Pediatr Surg* 16: 471–475.
17. Siddiqui AR (1985) 'Hepatobiliary System.' In: Siddiqui AR (eds), *Nuclear Medicine in Pediatrics*, pp 123–140. Yearbook Medical Publishers, Inc.
18. Howard ER (1983) 'Extrahepatic biliary atresia: a review of current management.' *Br J Surg* 70: 193–197.
19. Rosenthal L (1980) 'Hepatobiliary Imaging.' Abdomen Alavi A, Arger PH (eds), pp. 49–72. New York: Grune & Stratton.
20. Shih W-J, Duff D, Mostowycz L (1988) 'Bile leakage accumulating in the gallbladder bed mimicking normal visualization of gallbladder in Technetium-99m Disida hepatobiliary imaging.' *Nucl Med Radiol Ser* 222–223.
21. Samuels LD, Grosfeld JL (1970) Serial scans of liver regeneration after hepatectomy in children.' *Surg. Gynecol. Obstet.* 131: 453–457.
22. Lin TY, Lee CS, Chen CC et al. (1979) 'Regeneration of human liver after hepatic lobectomy studied by repeated liver scanning and repeated needle biopsy.' *Ann Surg* 190: 48–53.
23. Vajrabukka T, Bloom AL, Sussman M et al. (1975) 'Postoperative problems and management after hepatic resection for blunt injury to the liver.' *Br J Surg.* 62: 189.
24. Krom R, Wiesner R, Pettke S, et al. (1989) 'The first 100 liver transplantations at the Mayo Clinic.' *Mayo Clin Proc* 64: 84–94.
25. Herry JY, Bissot P, Le Jeune JJ et al. (1980) 'Evaluation of liver transplant by ^{99m}Tc dimethyl-IDA scintigraphy.' *J Nucl Med* 21: 657–659.
26. Loken MK, Ascher NL, Mikhail SJ et al. (1987) 'Functional imaging of liver transplants.' *J Nucl Med* 24: 98.
27. Hall TR, Hawkins RA, Gambhir SS et al. (1987) 'Quantitative dynamic HIDA studies for evaluation of pediatric liver transplant function.' *Radiol* 165(P): 211.
28. Cronin EB, Ichise M, Summerville DA, et al. (1987) 'Hepatobiliary imaging in pediatric liver transplantation.' *Radiol* 165(P): 212.
29. Taormina V, McLean GK (1985) 'Chronic bile peritonitis with progressive bile ascites: a complication of percutaneous biliary drainage.' *Cardiovasc Intervent Radiol* 8(2): 103–105.
30. McGuigan J (1987) 'Peptic ulcer, Chapter 235 in Harrison's principles of internal medicine.' 11th ed. Braunwald et al. (eds), McGraw Hill Book Co., New York, p 1249.

5. Measurements of liver haemodynamics

DUNCAN ACKERY

Introduction

The fundamental principle underlying static radionuclide imaging procedures is that concentration of radioactivity takes place following the intravenous administration of a radiopharmaceutical. Uptake of a radiopharmaceutical depends upon regional cardiac output to the tissue under examination and on the extraction efficiency by the tissue. The product of these two factors is often called the Effective Organ Blood Flow. It is important to make the distinction between blood perfusion, that is the pattern of relative blood flow to the organ, and the absolute blood flow, measured in millilitres of blood per minute flowing in a unit mass of tissue. Most radionuclide clinical investigations require only the perfusion pattern to be measured and absolute measurements are seldom necessary. Also it is usually difficult and unnecessary to measure the extraction efficiency of the radiopharmaceutical.

In this chapter the methods of radionuclide measurement of liver blood flow will be discussed, with a description of the different techniques and radiopharmaceuticals available and the mathematical models underlying the principles of the methods. The analytical procedures for measuring fractionated perfusion will be reviewed with an indication of their clinical utility.

Radionuclide imaging of the liver has been available for more than two decades and has been used primarily for the identification of space occupying lesions. In recent years other imaging methods have been shown to have better resolution for the detection of focal lesions together with certain other diagnostic advantages; consequently in many hospitals the number of radionuclide studies has declined in favour of ultrasound, computed tomography and nuclear magnetic resonance.

It has been appreciated for some time that early images of the haemodynamic phase can add discrimination to radionuclide liver imaging, [1–4], and this has stimulated an increasing interest in the use of "first pass" radionuclide techniques. From these the degree of vascularity of focal lesions compared to that of normal hepatic parenchymal tissue can be demonstrated. More recently the use of digital methods with fast on-line acquisition of data has

H.J. Biersack and P.H. Cox (eds.), Nuclear Medicine in Gasteroenterology, 69–85.

permitted quantitative analysis of activity-time curves. Using mathematical modelling [5] these have been used for the estimation of total hepatic blood flow, for assessment of the proportionate afferent flow between the hepatic artery and portal vein and for measuring the pattern of microsphere distribution following direct hepatic arterial injection.

Alterations in organ perfusion or function may give useful diagnostic information. However changes in hepatic blood flow, and in the fractional contribution from the hepatic artery and portal vein, occur also under different physiological conditions [6]. A four fold variation in liver blood flow under normal conditions in animals has been attributed to irregularity of flow through the sinusoids, and this is supported by changes in sinusoid flow observed directly in the trans-illuminated liver. Both upright posture and exercise cause a relative drop in the cardiac output to the liver. Flow in the mesenteric circulation increases following feeding both in absolute terms and in the proportion perfusing the liver, which rises from 59% to 76% of total flow between fasting and taking a standard meal [7]. It is recommended that clinical studies are undertaken in the supine position, after a period of rest, and after overnight fasting.

First pass techniques

Most radionuclide techniques for observing hepatic haemodynamics use a gamma scintillation camera. This may be placed above the supine subject with the field of view including the heart, lung bases, major abdominal blood vessels, liver, spleen and kidneys. Alternatively the camera is positioned under the bed allowing better visualization of the posterior hepatic right lobe and kidneys. The radiopharmaceutical is drawn up in a small (under 0.5 ml) volume and administered rapidly as a bolus into an antecubital vein. This is followed by a saline flush which assists the rapid transit of the activity through the heart and into the systemic circulation. The most convenient method of delivery is through an in-dwelling venous line using a three-way tap. Special syringes are available which permit the radiopharmaceutical and saline to be drawn up in a single syringe without a tap.

Data is acquired onto an on-line computer typically in a 64×64 pixel frame format in two phases; for example using 0.5 s intervals for 40 s (80 images) followed by 15 s frames (60 images). The first stage acquires data on the first pass radiocolloid into the liver from which the arterial and portal components of hepatic blood flow are determined. The second stage measures the rate of colloid clearance and gives an index of total hepatic reticuloendothelial flow. Summation of early frames allows anatomical definition of structures for region of interest analysis.

Examination of the time-activity curve over the aortic region of interest defines the technical acceptability of the study. A bolus is considered acceptable if the leading width of the half maximun of the curve is equal to or less than 15% of the time (seconds) to peak [8]. Alternatively the full width at half maximum value for the peak activity in the heart region of interest should not exceed 12.5 s [9].

Radiopharmaceuticals

Reported studies of dynamic liver blood flow have used either the tracer pertechnetate (^{99m}Tc) or colloids labelled with ^{99m}Tc which are taken up by the liver parenchyma. During the first pass of pertechnetate into the liver the counts rise to a peak and then drop as activity is removed in the venous outflow. On the other hand radio-labelled colloid is extracted by reticulo-endothelial macrophages found primarily in the liver (Kuppfer cells) and to a lesser extent in the spleen and bone marrow. In the healthy liver more than 80% of activity is cleared from the circulation during the first pass. Extraction varies with colloid particles of different sizes, and is impaired in liver disease. The time-activity curve over the liver for radiocolloid does not show a decline due to venous washout and continues to rise. Radiocolloid arriving at the liver in the portal vein is comprised almost entirely from that which has passed through the mesenteric circulation. Activity which has taken the splenic route will be removed by the spleen. The proportional portal flow when measured with radiocolloid is therefore often referred to as the 'mesenteric fraction.'

Following injection of radiocolloid clearance takes place according to a three compartment model [10]. Applying this model to in vivo measurements with a gamma camera Reske [11] designates the first compartment to the particle distribution pool, the second to particle turnover at the macrophage membrane and the third to the intracellular space of Kuppfer cells, reflecting the time course of phagocytosis and metabolic degradation. The use of pertechnetate and technetium labelled colloid in the same subject permits assessment of the relative contributions of extrahepatic and intrahepatic circulations and that of reticulendothelial extraction [12].

Radiocolloid has certain advantages over pertechnetate for measuring the arterial and portal components of flow [13]. The clearance of most of the activity from the splenic circulation improves the temporal separation of arterial and portal phases as the mesenteric pathways are longer than those for the spleen. Also the uptake of colloid by the liver simplifies the choice of appropriate regions of interest, and following the haemodynamic study a conventional hepatic image can be obtained.

Other radiopharmaceuticals that may be used to investigate hepatic haemodynamics are labelled microspheres, which are extracted in the first passage through a capillary bed and thus require direct arterial or portal injection, and radioactive xenon, a poorly soluble gas which is washed out from the liver at a rate proportional to the hepatic blood flow.

Mathematical models

Direct measurement of clearance by arterial and venous sampling

These techniques measure the clearance of dyes, radioactive tracers and other substances by direct sampling from the hepatic veins. Using the Fick principle the hepatic blood flow, Q_H, can be determined as follows:

$$Q_H = \frac{R}{C_i - C_0}$$

Where:

R = total amount of tracer removed by the liver per minute
C_i = concentration of tracer per millilitre entering the liver
C_0 = concentration of tracer per millilitre exiting the liver

The assumptions made are that (a) material is removed only by the liver, (b) that single hepatic vein sampling represents total venous efflux, (c) that extrahepatic shunting does not take place, and (d) that extraction efficiency remains constant.

Comment. This technique is not generally applicable to clinical study as it requires hepatic vein sampling with constant infusion of tracer.

Clearance of tracer (e.g., colloid) from peripheral blood

For a single compartment system the first order kinetics are given by:

$$C(t) = C(0) \cdot e^{-kt}$$

where

$C(t)$ = the concentration of colloid at time t
$C(0)$ = the initial concentration of colloid
k = the rate constant of disappearance

The total hepatic blood flow, Q_H, is given by:

$$Q_H = \frac{V}{E}(k)$$

Where

V = the blood volume (ml)

E = the extraction efficiency of colloid by the liver

The effective liver blood flow, $E \cdot Q_H$, can therefore be expressed as

$$E \cdot Q_H = V \cdot k$$

The assumptions here are that (a) single compartment kinetics pertain, (b) extraction efficiency remains constant and is not affected by disease, (c) extrahepatic shunting does not occur, (d) tracer is extracted only by the liver, and (e) saturation of the uptake mechanism involved in the extraction does not occur [14].

Comment. If the tracer is extracted by more than one site, e.g., radiocolloid, then k, the extraction rate, is the sum of all the individual extraction rates for each site. Colloid of different particle size will have different extraction efficiencies.

Determination of hepatic blood flow by in vivo counting without blood sampling

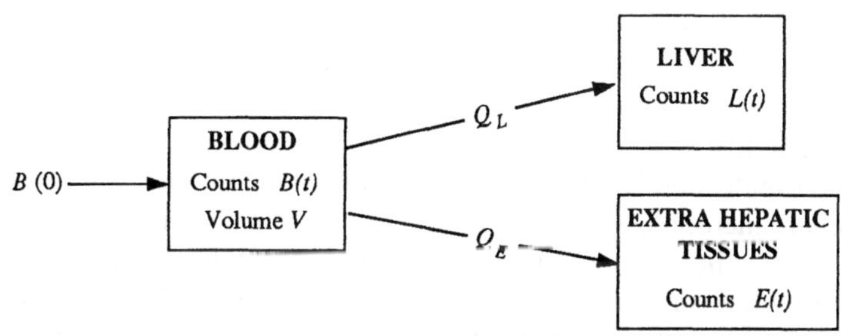

The differential equations which follow describe the varying activities (counts per second) which occur in the blood, $B(t)$, liver, $L(t)$, and in extrahepatic sites, $E(t)$, following administration of colloid activity, $B(0)$. Q_L and Q_E are the effective perfusion rates (ml/min) of the liver and extrahepatic sites, respectively. V is the blood volume (ml).

Then:

$$B(t) = B(0) \cdot e^{-(Q_L+Q_E)t/V}$$

$$L(t) = \frac{B(0) \cdot Q_L}{Q_L + Q_E} \cdot (1 - e^{-(Q_L+Q_E)t/V})$$

$$E(t) = \frac{B(0) \cdot Q_E}{Q_L + Q_E} \cdot (1 - e^{-(Q_L+Q_E)t/V})$$

It follows that:

$$\frac{L(\infty)}{E(\infty)} = \frac{Q_L}{Q_E}$$

where $L(\infty)$ = plateau liver counts, and $E(\infty)$ = plateau extrahepatic counts, therefore

$$\frac{\text{(Counts in the liver)}}{\text{(Counts in all other tissues)}} = \frac{Q_L}{Q_E}$$

also:

$$\frac{Q_L + Q_E}{V} = \begin{array}{l}\text{rate constant of colloid clearance}\\ \text{(e.g., measured over the liver)}\end{array}$$

V = plasma volume (measured, or calculated from nomogram)

therefore Q_L can be calculated as the total effective perfusion rate of the liver in ml/min.

Comment. In this procedure it is assumed that radiocolloid mixes homogeneously and rapidly and is removed from the vascular compartment by reticuloendothelial macrophages, primarily in the liver but also in the spleen, bone marrow and other tissues. The model does not take account of colloid extracted by the spleen, but a correction can be made for this [7]. It should be noted that the rate of colloid clearance at all sites is equal to the sum of the rates at each site [16].

This approach can be used to analyse tracer uptake by the liver and spleen after correction for organ depth and tissue background. Measurements of hepatic and splenic uptake rate give good predictive accuracy in a number of clinical disorders [16].

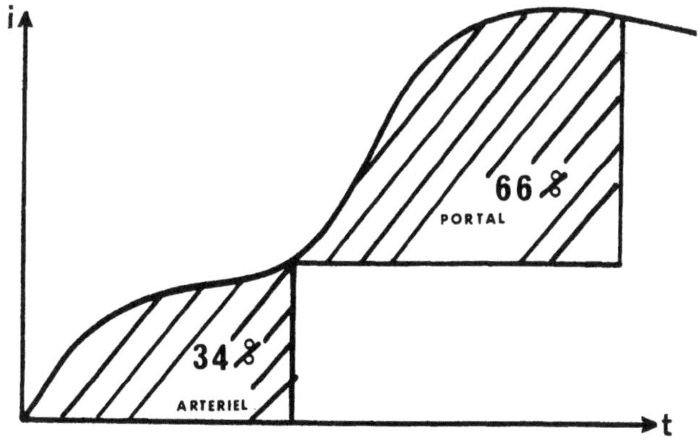

Figure 1. Biphasic distribution of acitivity in the normal liver following rapid bolus injection of radiopharmaceutical [20].

Measurement of the relative hepatic artery and portal vein perfusion

These measurements may be made by direct administration of radiopharmaceutical into the hepatic artery by catheterization, or into the portal system via percutaneous splenic puncture, cannulization of the umbilical vein, or direct injection of microspheres into the portal vein [17, 18]. Such techniques are invasive and do not measure the fractionated perfusion. Animal studies of relative cardiac output by each route can be carried out using intracardiac injection of labelled microspheres with arterial sampling [19].

Alternatively non-invasive studies can be carried out using pertechnetate or labelled colloid. Following rapid bolus injection of radiopharmaceutical into a peripheral vein activity passes through the heart and into the systemic circulation. Activity appears in the liver firstly through the hepatic artery, followed by that arriving in the portal venous system. Different analytical methods have been described to measure the proportionate flow via the two routes. This principle was first investigated by Taplin [1] and later by Biersack, who calculated the relative arterial and portal contributions to flow from the areas under biphasic time-activity curve [20, 21]. This is done by visual inspection and identification of the inflection point on the curve and assuming that this represents the beginning of the portal phase of the curve. The end point of portal phase is given when a plateau is reached (Fig. 1).

Hepatic Perfusion Index (HPI)

In this method the gradient of the time activity curve is measured for the two components of flow. A rapid bolus (0.2–0.5 ml) of activity is administered. Either pertechnetate or technetium colloid may be used. Rapid gamma camera images are acquired to include the heart, liver, lung bases spleen and kidneys. Data are stored for 100 s following injection. Sequential frames of the study are added and regions of interest drawn for the left ventricle,

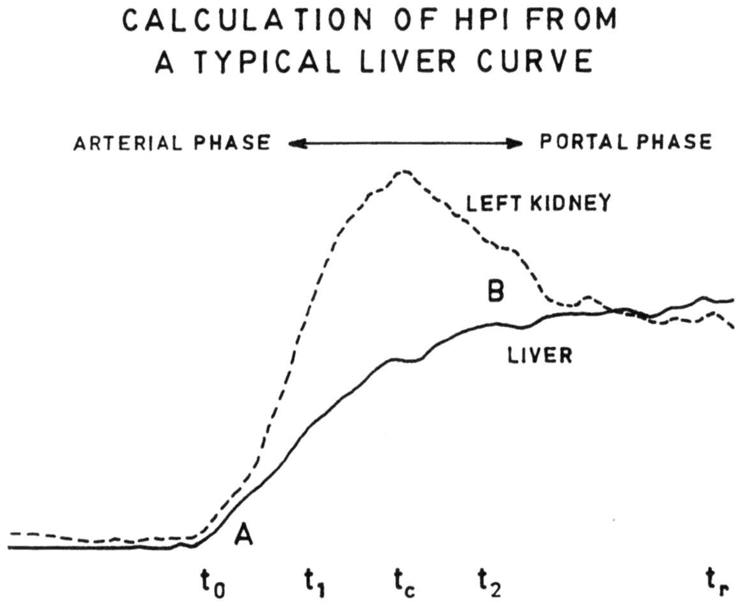

Figure 2. Calculation of HPI from a typical liver curve ($t_1 = t_0 + 7$ s, $t_2 = t_c + 7$ s). Slopes of lines A and B represent anterial and portal venous components of hepatic flow, respectively.

right hepatic lobe, right lung, spleen and kidneys. Particular care is taken to avoid major vessels and other regions of high background within the chosen region for the liver. Time-activity curves are plotted for the six regions.

The detailed analysis is described by Sarper [22]. Initially pulmonary activity is subtracted from the hepatic curve. Two time points are identified on the hepatic curve; the first, t_0, represents the earliest arrival of hepatic arterial activity into the liver, and the second, t_c, the time of maximal activity in the kidney. The curve following this time assumed to be due to the portal contribution. The average arterial slope is derived from $t_0 + 7$ s and the portal phase slope from $t_c + 7$ s (Fig. 2).

This method of analysis was later refined [23] to correct for the error which results in the early part of the portal phase due to washout of activity from the arterial component. This is done by estimating hepatic washout from the downslope of the splenic time-activity curve following the peak. The HPI is then calculated as the true portal component divided by the sum of the arterial and true portal components.

A variation on the analysis is described by Parkin [13]. These workers chose to acquire data from the posterior aspect so as to minimise the statistical inaccuracy in counts from the kidneys. Using a rapid bolus of technetium colloid data are acquired in 2 s (64 × 64) frames for one minute. Regions of interest are drawn over the liver and right kidney excluding regions of high background activity. The time of peak activity in the kidney is used to define the division between arterial and portal components of flow. The least squares gradient over the 8 s prior (arterial) and post (portal) of this time are taken and the PI calculated by the same ratio method as Sarper. Perkins [24] found the left kidney more reliable than the right for the HPI estimation with an upper limit for the normal value to be 0.37.

Nott [25] has shown that the Hepatic Perfusion Index can be reliably measured in small animals, and have validated the results by comparison with absolute flows using a microsphere technique [19].

Hepatic Arterial Ratio (HAR)

In this method patients are positioned supine over a large field of view gamma camera fitted with a high sensitivity collimator [26]. 150 MBq of technetium sulphur colloid are administered as a rapid intravenous bolus. Sequential posterior images are recorded for 5 min, initially at 2 s and later at 6 s intervals. Quantitative anterior and posterior images of the liver and spleen are recorded after maximum colloid clearance. After construction of activity-time curves from appropriate regions of interest the arterial component of hepatic flow is calculated by fitting the initial rise of the splenic curve to the hepatic curve. The arterial percentage hepatic flow is then calculated from the relative heights of the hepatic uptake curve and the arterial component at 5 min, applying a correction for the colloid removed from the portal circulation by the spleen. The rate of clearance of tracer is determined by a simple clearance index obtained from the cardiac blood pool activity at 30 and 120 s (Fig. 3).

Let:

C_a = height of the arterial curve at 5 min
C_h = height of the hepatic curve at 5 min
Q_s = splenic flow
Q_m = mesenteric (gut) flow (=portal (Q_p) – splenic flow)
Q_a = hepatic arterial flow
S = total activity in the spleen
L = total activity in the liver
R = fractional arterial flow (arterial/total)

Then:

$$\frac{C_a}{C_h} = \frac{Q_a}{Q_a + Q_m} \tag{1}$$

$$\frac{S}{L} = \frac{Q_s}{Q_a + Q_m}$$

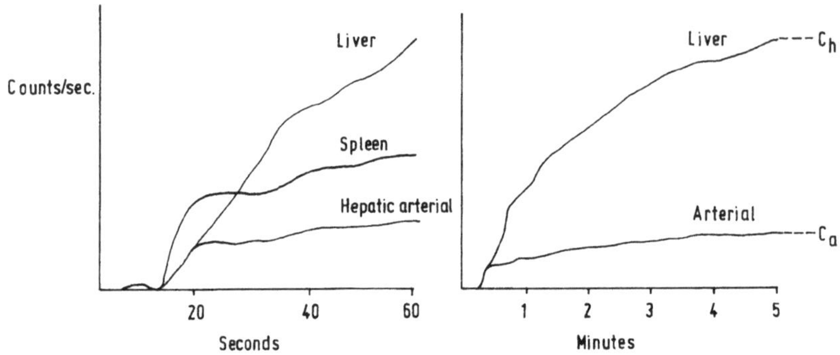

Figure 3. **Curves showing the time course of activity in the hepatic artery, spleen and liver, used to calculate the Hepatic Arterial Ratio [26].**

$$1 + \frac{S}{L} = \frac{Q_a + Q_s + Q_m}{Q_a + Q_m} = \frac{Q_a + Q_p}{Q_a + Q_m} \qquad (2)$$

Therefore from (1) and (2):

$$\frac{Q_a}{Q_a + Q_p} = \frac{C_a}{C_h(1 + S/L)}$$

and

$$R = \frac{C_a L}{C_h(L + S)}$$

Mesenteric Fraction (*MF*)

Using this method the temporal separation of the two phases of hepatic perfusion is given by comparison with activity in the cardiac blood pool, the spleen and left kidney [27]. The fractional perfusion is given by the relative heights of the liver activity-time curve as with the HAR method. The patient is positioned supine beneath a gamma camera and technetium colloid is administered in a bolus. Digital images are recorded in a 64 × 64 matrix at 0.5 s intervals for 40 s. Time activity curves are constructed from regions of interest for the heart, liver, spleen and left kidney.

To calculate the mesenteric fraction the end of the arterial phase in the liver is taken from a mean of the half peak decline value of cardiac activity, the plateau value of splenic activity, and the peak activity in the left kidney. Similarly three estimates are taken from the heart, spleen and left kidney to establish the termination of the portal phase.

Colloid arriving at the liver via the portal vein will comprise only that

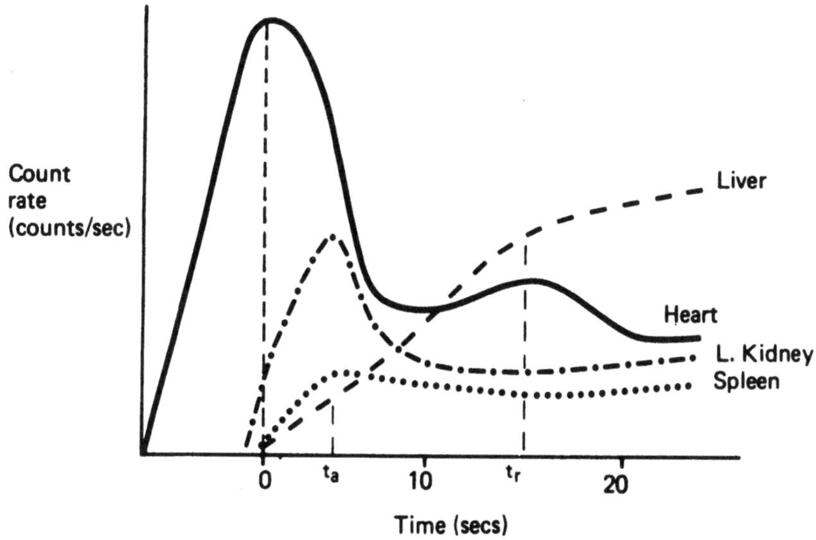

Figure 4. **Time course of activity in the heart, liver, spleen and left kidney following a bolus injection of activity, used to calculate the Messenteric Fraction [27].**

which has passed through the mesenteric circulation; activity in the splenic artery having been extracted in the spleen. Thus the mesenteric fraction of hepatic perfusion, MF, can be given by:

$$\mathrm{MF} = \frac{Q_m}{Q_a + Q_m} = 1 - \frac{L_a}{L_p}$$

where:

Q_m = mesenteric artery perfusion
Q_a = hepatic arterial perfusion
L_a = the activity in the liver at the end of the arterial phase
L_p = the activity in the liver at the end of the portal phase

The method has been validated by measurements in experimental animals following ligation of the hepatic artery and porta-caval anastomosis [17].

Comparison of different techniques and errors

The various methods of analysis used to determine the contribution of flow to the liver via the hepatic artery and portal vein use different assumptions and variables. Both physiological and physical factors may affect the result [28]. Considerable variations for the values of HPI are obtained depending on the number of data points used in the determination of the flow gradient or the number of smoothing operations to which the data are subjected. The use of an extracted tracer is recommended, and it is possible that the currently considered abnormal ranges for HPI may be poor indicators of the fractional flow.

In a study of the variability between different observers using the HPI analysis for patients with colorectal malignancy better agreement was obtained when the left renal peak activity was used to define the end of the arterial phase than the right [29].

In a comparison of three different methods of analysis (HPI, HAR, MF) Britten [30] acquired data from normal and patients with colorectal malignancy using the anterior protection. Reasonably good correlation was shown between HPI and MF, but that between HAR and the other indices was less good. Paired results for different regions of the liver analysed by each method show a significant variation for HPI and MF, but not for HAR [31]. In addition calculation of the index (1-MF) for each pixel in the hepatic region gives a parametric image which improves the choice of selection of regions of interest and the separation of hepatic activity from that in overlying tissues.

Experimental and clinical studies of liver flow

Hepatic malignancy

Early identification of hepatic metastases is important for the correct management of patients with malignancy. Recognition of small focal lesions within the normal hepatic parenchyma depends upon the spatial resolution of the imaging method which is used. Conventional radionuclide colloid imaging is unable to detect lesions of less than about 2 cms, particularly if they lie deep in the liver. The resolution of ultrasound, computed tomography and nuclear magnetic resonance is somewhat better but still has a finite limitation of detection. Metastases in the liver are primarily nourished by hepatic arterial blood and so it has been proposed that the arterial component of flow might be proportionately increased with early metastatic involvement.

Leveson [32] using the Hepatic Perfusion Index has shown significant increase in the fraction of hepatic arterial blood to the liver in those patients with positive laparotomy even when no metastases were visible by conventional radionuclide imaging. In a later publication [33] the HPI was found to be increased in patients with occult metastatic disease, and these authors claim a sensitivity of 96% and a specificity of 72% as judged by the finding of metastases one year later. The same group investigated 150 patients with primary carcinoma of the gastrointestinal tract and showed that the Hepatic Perfusion Index was elevated in 94% of those with liver metastases at laparotomy and in 87% of those with occult metastases which became overt within three years [34].

It remains unclear whether the apparent rise in hepatic arterial contribution is not in fact due to reduction in portal flow to the liver. It has been shown that metastases from colorectal carcinoma which spread to liver via the portal vein preferentially block the portal inflow to the liver [35]. Nott [36] used a rat model to investigate this. Micrometastases were induced in animals by intraportal inoculaticn of Walker carcinosarcoma cells. The Hepatic Perfusicn Index was measured at 2, 4 and 6 days and showed a significant rise. However hepatic arterial flow, as determined by intra-ventricular injection of labelled microspheres, did not alter and portal flow was proportionately reduced. On the other hand using the MF technique it has been shown [37] that the absolute arterial flow increases proportionately with the amount of hepatic replacement by colorectal metastases.

Cirrhosis

A number of different methods have been described to measure the severity of diffuse liver disease from quantification of the radionuclide colloid image. Initially this was done by simply summing of the counts in the liver and spleen and expressing a ratio of activity. Increasingly impaired hepatic function was indicated by a reduced liver/spleen ratio. Better discrimination was shown by measuring vascular clearance of colloid by means of a probe placed over the aortic arch [38] and combining this with scintillation camera studies of hepatic and splenic uptake.Using the mathematical model described on pages 73 and 74 Miller[39] showed a significant reduction in liver uptake rate of colloid in cirrhotic patients compared to normals, but not for those with alcoholic hepatitis. This is supported by other studies [26, 40–42]. In animals there is a close correlation between rise in portal pressure and shunting, measured by the diversion away from the liver of radiolabelled microspheres injected into the portal vein. Beta-1-blocking drugs reduce liver blood flow in normal rats and have a variable effect in cirrhotic animals, probably reflecting the balance between changes in cardiac output and hepatic perfusion [43]. Measuring the mesenteric fraction in humans McLaren [44] showed some overlap of cirrhotics with a healthy group, but obtained better discrimination when data for mesenteric fraction were combined with liver/ spleen ratios measured as a geometric mean. Patients who had recently bled from varices showed a greater reduction in values. Holbrook [45] concluded that patients who have bled and have a low initial value of mesenteric fraction have a significant risk of rebleeding despite active prophylactic treatment. In an attempt to estimate the degree of shunting of the portal circulation Mairing [46] has undertaken studies in cirrhotic patients using simultaneous technetium colloid and pertechnetate. The results permit calculation of a ‘trapping index’ which they have found to be a reliable indicator of the risk of bleeding.

Liver graft rejection

Measurement of the portal contribution to hepatic flow can be used to monitor the status of patients following hepatic transplantation [47]. In general the transplant has a lower portal contribution than normal liver. The level drops below 55% with acute rejection but remains above this value in post-transplant hepatitis.

Portal occlusion and mesenteric ischaemia

A good correlation is shown between measurement of mesenteric fraction and the degree of portal shunting using direct portal injection of labelled microspheres in experimental animals in which portal hypertension has been produced by progressive portal vein ligation [48]. The results suggest that non-invasive radionuclide methods could be used clinically for assessing the extent of portal flow prior to hepatic arterial embolization [49], and for estimating the degree of mesenteric ischaemia in patients with symptoms of intestinal ischaemia.

Hepatic regeneration following surgery

After partial resection in children and young adults the remaining liver grows rapidly to its original mass by about 3–4 weeks. Measurements of hepatic blood flow with radiocolloid has been used to show the haemodynamic changes after resection in animals [15].

Other methods of measuring hepatic blood flow

Radioactive gaseous tracers

The clearance of xenon-133 from hepatic parenchyma has been used to measure liver blood flow in animals and patients [50, 51]. The gas is dissolved in saline solution and administered directly into the hepatic artery or portal circulation (via dilatation and cannulation of the obliterated umbilical vein). This is followed by continous measurement by a collimated probe or gamma camera to give the time-activity clearance curve. This is a double exponential function the fast component of which represents blood flow.

Ultrasound Doppler techniques

The combination of B mode ultrasound imaging with pulsed Doppler has recently been shown to be promising in the measurement of portal blood flow [52]. The procedure is difficult and time-consuming, but both a qualitative and quantitative assessment of flow in the portal vein is possible and this has been used for the assessment of patency of, and direction of flow in, portosystemic shunts, and in the identification of portal vein thrombosis. Visualization of the larger hepatic arteries and veins is also possible and has been used in patients with cirrhosis, in the assessment of flow prior to hepatic transplant, and in the Budd–Chiari syndrome. A number of technical difficulties still exist, but as these are overcome ultrasound could have an important role in both the qualitative and quantitative assessment of liver blood flow.

Radiological contrast angiography

Angiography through the coeliac or hepatic arteries gives a qualitative impression of hepatic arterial perfusion. It has been shown to have value when combined with radionuclide angiography for giving greater discrimination in the detection of a variety of hepatic conditions [53].

Conclusion

Nuclear medicine provides simple non-invasive procedures for measuring hepatic haemodynamics. Toxicity and radiation absorbed doses are low. So far these procedures have shown clinical promise rather than any routine place in the management of hepatic disorders. The relative arterial component has been shown to increase in cirrhosis, hepatic metastases and mesenteric ischaemia. There is therefore a possible place for the monitoring of patients with cirrhosis or following hepatic transplantation, for assessing of adequacy of perfusion prior to embolization, and in the early detection of hepatic secondary malignancy.

References

1. Taplin GV (1971) 'Dynamic studies of liver function with radioisotopes.' In: *Dynamic Studies with Radioisotopes in Medicine*. pp. 373–392. New York: Unipub Inc.
2. Waxman AD, Apau R, Siemsen JK (1972) 'Rapid sequential liver imaging.' *J Nucl Med* 13: 522–524.
3. Witek JT, Spencer RP (1974) 'Clinical correlation of hepatic flow studies.' *J Nucl Med* 16: 71–72.
4. Houston AS, MacLeod MA (1980) 'Processing of liver dynamic studies with technetium-labelled sulphur colloid.' *Brit J Radiol* 53: 87–92.
5. Magrini A, Izzo G, Guerrisi M, et al. (1985) 'A new approach to non-invasive quantitative study of hepatic haemodynamics uisng radiocolloids in vivo.' *Clin Phys Physiol Meas* 6: 179–204.
6. Bradley SE (1949) 'Variations in hepatic blood flow in man during health and disease.' *New Eng J Med* 240: 456–461.
7. Walmsley BH, Fleming JS, Ackery DM, Karran SJ (1987) 'Noninvasive assessment of absolute values of hepatic haemodynamics using radiocolloid scintigraphy.' *Nucl Med Commun* 8: 613–621.
8. Boyd RO, Stadalnik RC, Barnett CA, Hines HH (1978) 'Quantitative hepatic sciintiangiography.' *Clin Nuc Med* 3: 478–484.

9. Fleming JS, Ackery DM, Walmsley BH, Karran SJ (1983). 'Scintigraphic estimation of arterial and portal blood supplies to the liver.' *J Nucl Med* 24: 1108–1113.
10. Dobson EL, Jones HB (1952) 'The behaviour of intravenously injected particulate material; the rate of disappearance from the blood stream as a measure of liver blood flow.' *Acta Med Scand* 144: Suppl 273.
11. Reske SN, Vyska K, Feinendegen LE (1981) 'In vivo assessment of phagocytic properties of Kupffer cells.' *J Nucl Med* 22: 405–410.
12. Izzo G, Diluzio S, Guerris M, Favella A, Magrini A. (1983) 'On the interpretation of the early part of the liver time activity curve: double tracer experiment.' *Eur J Nucl Med* 8: 101–104.
13. Parkin A, Robinson PJ, Baxter P, et al. (1983) 'Liver perfusion scintigraphy – method, normal range and laparotomy correlation in 100 patients.' *Nucl Med Commun* 4: 395–402.
14. Bradley EL (1974) 'Measurement of hepatic blood flow in man.' *Surgery* 75: 783–789.
15. Karran SJ, Eagles CJ, Fleming JS, Ackery DM (1979) 'In vivo measurement of liver perfusion in the normal and partially hepatectomized rat using ^{99m}Tc sulfur colloid.' *J Nucl Med* 20: 26–31.
16. Rutland MD (1984) 'An analysis of the uptake of ^{99m}Tc-sulphur colloid by the liver and spleen.' *Nucl Med Commun* 5: 593–602.
17. Kashiwagi T, Kamada T, Abe H (1974) 'Dynamic studies on the portal hemodynamics by scintiphotosplenoportography: the visualization of portal venous system using ^{99m}Tc.' *Gastroent* 67: 668–673.
18. Gross G, Goldberg HI, Shames DM (1976) 'A new approach to evaluating hepatic blood flow in the presence of intrahepatic portal systemic shunting.' *Invest Radiol* 11: 146–149.
19. McDevitt DG, Nies AS (1976) 'Simultaneous measurement of cardiac output and its distribution with microspheres in the rat.' *Cardiovasc Res* 10: 494–498.
20. Biersack HJ, Thelen M, Schulz D, Knopp R, Dahlem R, Schmidt R, Winkler C (1977) 'Die sequenttielle hepatospleno-szintigraphie zur quantitativen beurteilung der leberdurchblutung.' *Fortschr Rontgenstr* 126: 47–52.
21. Biersack HJ, Torres J, Thelen M, Molon O, Winkler C (1981) 'Determination of liver and spleen perfusion by quantitative sequential scintigraphy: results in normal subjects and in patients with portal hypertension.' *Clin Nucl Med* 6: 218–220.
22. Sarper R, Fajmam WA, Rypins EB, Henderson JM, Tarcan YA, Galambos JT, Warren WD (1981). 'A non-invasive method for measuring portal venous/total hepatic blood flow by hepatosplenic radionuclide angiography.' *Radiology* 141: 179–184.
23. Sarper R, Tarcan YA (1983) 'An improved method of estimating the portal venous fraction of total hepatic blood flow from computerized radionuclide angiography.' *Radiology* 147: 559–562.
24. Perkins AC, Whalley DR, Ballantyre KC, Hardcastle JD (1987) 'Reliability of the hepatic perfusion index for the detection of liver metastases.' *Nucl Med Commun* 8: 982–989.
25. Nott DM, Grime JS, Yates J, O'Driscoll MP, Baxter JN, Cooke TG, Jenkins SA (1987) 'A model of the hepatic perfusion index in the rat.' *Nucl Med Commun* 8: 990–994.
26. Wraight EP, Barber RW, Riston A (1982) 'Relative hepatic arterial and portal flow in liver scintigraphy.' *Nucl Med Commun* 3: 273–279.
27. Fleming JS, Humphries NLM, Karran SJ, Goddard BA, Ackery DM (1981) 'In vivo assessment of hepatic arterial and portal venous components of liver perfusion: Concise communication.' *J Nucl Med* 22: 18–21.
28. Tindale WB, Barber DC (1987) 'The effect of methodology and tracer identity on a non-invasive index of liver blood flow.' *Nucl Med Commun* 8: 973–981.
29. Whalley DR, Perkins AC, Ballantyre KC, Hardcastle JD (1987) 'Validity of the hepatic perfusion index for the detection of liver metastases.' *Nucl Med Commun* 8: 271–272.
30. Britten AJ, Fleming JS, Flowerdew ADS, Taylor I, Karran SJ, Ackery DM (1990) 'Regional indices of relative hepatic arterial perfusion from dynamic liver scintigraphy: the variability of indices and the use of parametric imaging.' *Nucl Med Commun* 11: 29–36.
31. Britten AJ, Fleming JS, Flowerdew ADS, Hunt TM, Taylor I, Karran SJ, Ackery DM

(1990) 'A comparison of three indices of relative hepatic perfusion derived from dynamic liver scintigraphy.' *Clin Phys Physiol Meas* 11: 45–51.

32. Leveson SH, Wiggins PA, Nausiru TA, Giles VR, Robinson PT, Parkin PA (1982) 'Improving the detection of hepatic metastases by the use of dynamic flow scintigraphy.' *Br J Cancer* 47: 719–721.
33. Leveson SH, Wiggins PA, Giles GR, Parkin A, Robinson PJ (1985) 'Deranged liver blood flow patterns in the detection of liver metastases.' *Br J Surg* 72: 128–130.
34. Cooke DA, Parkin A, Wiggins P, Robinson PJ, Giles GR (1987) Hepatic perfusion index and the evolution of liver metastases.' *Nucl Med Commun* 8: 970–974.
35. Mooney B, Grime JS, Taylor I, Critchley M (1983) 'Portal scanning for liver metastases in colorectal carcinoma.' *Clin Rad* 34: 657–659.
36. Nott DM, Grime JS, Yates J, Day DW, Baxter JN, Jenkins SA, Cooke TG (1987) 'Changes in the hepatic perfusion index during the growth and development of experimental micrometastases.' *Nucl Med Commun* 8: 995–1000.
37. Hunt TM, Flowerdew ADS, Britten AJ, Fleming JS, Karran SJ, Taylor I (1989) 'An association between haemodynamic parameters obtained by dynamic liver scintigraphy and percentage hepatic replacement with tumor.' *Brit J Cancer* 59: 410–414.
38. DeNardo SJ, Bell GB, DeNardo GL, Carretta RF, Scheibe PO, Imperato TJ, Jackson PE (1976) Diagnosis of cirrhosis and hepatitis by quantitative hepatic and other reticuloendothelial clearance rates.' *J Nucl Med* 17: 449–459.
39. Miller J, Diffey BL, Fleming JS (1979) Measurement of colloid clearance rate as an adjunct to static liver imaging.' *Eur J Nucl Med* 4: 1–5.
40. Ferguson WR, Laird JD, Cranley K (1981) 'Early dynamic studies as an adjunct to liver scintigraphy in the investigation of diffuse liver disease.' *J Nucl Med* 22: P88.
41. Leng B, O'Driscoll MP, Majeed FA, Grime JS, Critchley M (1987) 'Hepatic perfusion index in cirrhotic livers – investigation of imaging and analytical procedures.' *Nucl Med Commun* 8: 1001–1010.
42. Stewart C, Sakimura I, Siegel ME, Harley H, Lee K (1984) 'The hepatic-arterial/portal-venous scintiangiogram in alcoholic hepatitis.' *J Nucl Med* 25: P67.
43. McLaren M, Braye S, Fleming J, Karran S, Taylor I (1987) 'Changes in blood flow, portal pressure and shunting during the development of cirrhosis in response to beta-blockage.' *Gut* 28: 663–667.
44. McLaren MI, Fleming JS, Walmsley BH, Ackery DM, Taylor I, Karran SJ (1985) 'Dynamic liver scanning in cirrhosis.' *Br J Surg* 72: 394–396.
45. Holbrook AG, Burge D, Fleming JS, McLaren MI, Taylor I, Karran SJ (1987) 'Dynamic hepatic scintigraphy in the prediction of recurrent variceal bleeding.' *Br J Surg* 74: 527.
46. Mairiang EO, Parkin A, Robinson PJ, et al. (1986) 'Noninvasive indices of liver blood flow in patients with complications of cirrhosis.' *Nucl Med Commun* 7: 268–269.
47. Martin-Comin J, Mora J, Figueras J, Puchal R, Jaurrita E, Badosa F, Ramos M (1988) 'Calculation of portal contribution to hepatic blood flow with ^{99m}Tc-microcolloids: a non-invasive method to diagnose liver graft rejection.' *J Nucl Med* 29: 1776–1780.
48. Burge DM, Holbrook AG, Karran SJ (1987) 'Non invasive assessment of portosystemic shunting in extrahepatic portal hypertension in rats.' *J Paed Surg* 22: 211–214.
49. Flowerdew ADS, McClaren MI, Fleming JS, Britten AJ, Ackery DM, Birch SJ, Taylor I, Karran SJ (1987) 'Liver tumour blood flow and responses to arterial embolization measured by dynamic hepatic scintigraphy.' *Br J Cancer* 55: 269–273.
50. Smith A, Clarke MB (1976) 'The determination of hepatic blood flow in the rat using xenon-133.' *Int J Appl Rad Isotop* 27: 201–210.
51. Sherriff SB, Smart RC, Taylor I (1977) 'Clinical study of liver blood flow in man measured by ^{133}Xe clearance after portal vein injection.' *Gut* 18: 1027–1031.
52. Becker CD, Cooperberg PL (1988) 'Sonography of the hepatic vascular system.' *AJR* 150: 999–1005.
53. Waxman AD, Finck EJ, Siemsen JK (1974) 'Combined contrast and radionuclide angiography of the liver.' *Radiology* 113: 123–129.

6. Hepatic scintigraphy for evaluation of liver grafts

KLAUS F. GRATZ, OTMAR SCHOBER
and BURCKHARD RINGE

Introduction

The first attempt to replace a human liver was made at the University of Colorado, Denver, in 1963 by T.E. Starzl and colleagues. Several years later teams in Boston and Denver had developed techniques for liver transplantation in dogs. Until 1967 these efforts were followed only by consecutives failures in three American institutions. The first extended survival of a human recipient was achieved in 1967 by the group around T. E. Starzl in Pittsburgh, USA. In Europe, Calne at Cambridge University, UK, started his programme 1968 and Gütgemann in Bonn, West Germany, tried at first in Germany to transplant human livers. At time, more than 50 orthotopic transplantations were done every year at Hannover Medical School.

Non-metastatic primary liver tumors and end-stage cirrhosis of the liver are the main indications for liver transplantation at time (Table 1). Survival for both groups of indications is not satisfactory. Especially after transplantation for liver cirrhosis a longer survival time is expected, because this is a principally benign disease. Five-years survival rates of 50 to 70% [1] are reported. The survival rates after one year are always greater than 60% (70–80%, in children: 80–90%), if the operation is done in an elective patient status. In patients with cirrhosis success of transplantation is dependent on the degree of disease, secondary involvement of other organ systems and the degree of portal hypertension. More than 40% of the tumor patients survive one year. Tumor recurrencies are the major problem, more than perioperative complications or problems due to the transplanted organ.

A distinct group of indications are inborn errors of metabolism (oxalosis, homozygous familial hyper-cholesterolemia, Wilson's disease, alpha-1-antitrypsin deficiency, Niemann-Pick disease etc.), which have been treated effectively with liver replacement in singular cases [1, 2]. More experience is necessary to establish transplantation as routine therapy in these cases.

Beside conventional, cadaveric whole-liver transplantation techniques for the implantation of hepatic lobes have been developed using a caderveric or living donor [3]. Reduced-size transplantation means the implanting of a lobe into a recipient smaller than the donor. The dividing of a larger liver from

H.J. Biersack and P.H. Cox (eds), Nuclear Medicine in Gasteroenterology, 87–99

Table 1. Indications for Liver Transplantation (1975–1985)

Maligne liver tumors (38%)	Livercirrhosis (62%)
Hepatocellular carcinoma (18%)	Posthepatic cirrhosis (15.5%)
Bile duct carcinoma (8%)	Primary biliary cirrhosis (11%)
Cholangiocellular carcinoma (4.5%)	Bile duct atresia (7.5%)
Metastasis (4%)	Aethylic cirrhosis (5%)
Papillomatosis (0.5%)	Cholangitis (4.5%)
	Budd-Chiari-syndrome (3%)
Miscellaneous tumors (3%)	Miscellaneous (15.5%)

a cadaveric donor and implanting of the lobes into two different recipients is known as split-liver transplantation. Both techniques have had survival rates of about 80%. Like full-size transplants, transplanted liver lobes grow along with the patient. Because the liver generates fully, hepatic function of partial hepatectomized donors is not affected.

An indication for auxiliary liver transplantation is given by fulminant hepatic failure due to toxic involvement to wait up regeneration of the affected liver. Technical problems are to be solved, especially stable placement of the graft beside the original organ and sufficient blood supply. Newer developments are partial (lobal) auxiliary transplantation without complete removal of the involved liver.

Clinical problems

When clinical problems occur after liver transplantation, they have to be managed as early as possible, because these patients are seriously ill and therefore unable to compensate. An organ failure has disastrous consequences and is not managable for a long time without retransplantation. A retransplantation might only be successful if it is done in a good status of the patient [4]. In 10 to 20% of the liver transplantations a severe complication occurs. A retransplantation is necessary in about 50% of these cases. A list of complications after liver transplantation is given in Table 2.

Radionuclide methods

Cholescintigraphy

After intravenous injection, ^{99m}Tc-iminodiacetic (IDA) agents are carried in blood bound to proteins, mainly albumin. When the albumin-^{99m}Tc-IDA-complex arrives via the portal vein or the hepatic artery at the space of Dissé, dissociation takes place between albumin and ^{99m}Tc-IDA. From the sinusoids the tracer diffuses through the pores in the endothelial lining to bind to a specific membrane bound carrier analogous to bromosulfophthalein

Table 2. Complications after liver transplantation

1. Vascular insufficiency (bleeding, hematoma, infarction, portal vein or hepatic artery thrombosis, stenosis)
2. Graft rejection (acute, chronic)
3. Biliary complications (bile leakage, obstruction)
4. Hepatocellular damage
   (ischemic time, preservative damage, diseases of the donor, drug toxicity (cyclosporine A, azathioprine))
5. Infection (diffuse (viral, fungal, bacterial); circumscribed (abscess/necrosis))
6. Recurrence of primary disease
7. Cerebrovascular accidents, pulmonary embolism etc.

and bilirubin uptake. The degree of hepatic uptake in vivo depends on the structural configuration of the IDA compound, the level and strength of protein binding, the presence in serum of other competing organic anions and the functional integrity of the hepatocytes.

Once inside the hepatocyte, the tracer may be bound by various enzymes and/or undergo metabolism. The excretion of tracer into intrahepatic bile ducts is assumed to occur via active transport. From the intrahepatic bile ducts the tracer is taken to the extrahepatic bile via choleresis. From here it may enter the duodenum directly or be stored in the gallbladder for later release. Once inside the intestines, IDA does not enter the enterohepatic circulation and is thus effectively removed from the system of interest.

Therefore it is possible to differentiate time dependent the arterial inflow, the parenchymal extraction rate, the transport in the bile duct system, the storage function of the gallbladder and the transport in the intestine.

For sequence scintigraphy 400 MBq ^{99m}Tc-IDA-derivate is injected as bolus and the arterial inflow is documented by a gammacamera in the ventral view. This view is prefered because the blood flow into the abdominal organs is discriminated best by this technique. In a dedicated computer system 0.5 s frames are stored. For the analogous display we use 2 s intervals. In the parenchymal phase 10 minutes after injection the liver region is imaged in 4 views (500 000 cts/image; anterior, right anterior oblique, right lateral and posterior view). The excretion of the tracer is documented every 5 to 10 min (ventral view, physical decay corrected time equivalent to the image of the parenchymal phase). Since the gallbladder is resected in transplanted patients, gallbladder activity is not expected.

In case of searching a bile leakage it is necessary to scintigraphy as long as the major part of the activity (labeled bile) has left the liver to exclude intrahepatic bile leaks. For detecting stationary bile depots within the bowel loops the way of intraluminal intestinal activity has to be followed up by further images.

Quantitative radionuclide evaluations offer an objective approach to changes in liver function after transplantation [5]. Compartimental and non-compartimental models have been proposed and proven in a small number of patients [25]. Although liver uptake and excretion of hepatobiliary agents follow nonlinear Michaelis–Menten kinetics, for tracer studies the exchange between compartments can be assumed to be first order as long as the tracer itself does not cause a significant degree of saturation of carriers and/or storage sites. In liver function studies only trace amounts of IDA, which are insufficient to saturate the system, are being used. Bilirubin competes with IDA for carrier transport. The use of such indicators of hepatocyte clearance is deficient because it does not take into account the volume of distribution of the test substance and nonhepatic routes of excretion dependent on total blood bilirubin level [6].

Today, only numerical descriptions of maximum uptake and effective half-time are available. Laboratory parameters give superior and distinct information compared with tracer kinetics. Nevertheless, DeJonge et al. [7] found in animal experiments a significant increase in $t_{1/2}$ but no change in t_{max}. $T_{1/2}$ became abnormal before histological evidence of rejection was evident. Normal excretion half-time of DISIDA is given as not more than 25 min (Taylor et al. 1986: 21 ± 2 min, Krishnamurthy et al. 1989: 18.8 ± 2.5 min) [8, 9]. In clinical routine a high frequency of examinations is needed to establish IDA tracer kinetics in the diagnostic of rejection. Since prolongation of $t_{1/2}$ is not specific for rejection, this parameter does not help to confirm the diagnosis.

Perfusion scintigraphy

Intravenously injected millimicrospheres (MMS) labelled with ^{99m}Tc are mainly removed by the Kupffer cells of the liver. It is a three-step interaction with macrophages: (1) attachment at the receptor of the macrophage membrane, (2) phagocytosis and (3) intracellular degration of the material. These intracellular deposited millimicrospheres are particles ($0.5–2.0 \times 10^{-6}$ m) composed of a stannous-cloride heat-denatured albumin complex.

The attachment of ^{99m}Tc-labelled colloids at the specific receptor sites of the membrane is flow-dependent and only slightly changed by alterations in reticuloendothelial system (RES) function [10].

Total hepatic nutritive perfusion may be estimated by regional hepatic uptake or by colloidal clearance techniques, using a multicompartment model [11]. Plasma disappearance and hepatic uptake are measured by external counting and blood flow is calculated from the elimination rate constants for colloid clearance. Under the condition and constraints of a two-compartment model liver perfusion is only related to extraction ratio, the blood volume (theoretically about 70 ml/kg body weight) and inversely proportional to the half-time ($t_{1/2}$) of the clearance process [12]. In liver-transplanted patients blood volume changes with the clinical situation, which has to be taken into account for every estimation. The extraction ratio is not known in transplanted organs and differs surely from that known in normals or patients with cirrhosis.

Compared intra-individually there is an excellent reproducibility of measurements of the invasion constant k_1 into the liver and hepatic uptake. Therefore it is sufficient for clinical use to calculate only k_1 as k-value ($k = 1/t_{1/2}\ \mathrm{min}^{-1}$, normally $> 1/\mathrm{min}^{-1}$) with former measurements in the same individual and without sophisticated compartimental modeling.

After intravenous injection of 30–40 MBq ^{99m}Tc-MMS uptake is documented by a gammacamera in the ventral view over a period of 15 min in 60 frames (15 s/frame). The region of interest (ROI) is delineated around the whole liver. The uptake curve over time is fitted exponentially by the 'least mean square' method estimating the k-value automatically.

After colloid sequence scintigraphy a quantitative liver-spleen scan should be done to estimate liver- and spleen volume and to detect defects as recurrencies, metastases, abscesses or subcapsular hematomas. The liver region is imaged in 4 views (500000 cts/image; anterior, right anterior oblique, right lateral and posterior view). An ultrasound examination is routinely done, if a defect is suspected (Figs 1, 2).

In addition, for measuring the arterial and venous fraction of the hepatic blood flow a bolus injection of 400 MBq ^{99m}Tc-DTPA in maximal 0.5 ml volume followed by 20 ml saline infusion is used [13]. For radionuclide angiography 100 frames (0.5 s/frame) are stored in a computer system. For the analogous display we use 2 s intervals. The arterial inflow is documented by the gammacamera in the ventral view. This view is prefered because the blood flow into the abdominal organs and big vessels is discriminated best by this technique. Representative regions of exclusively arterial perfused organs (spleen or kidney) and of the liver parenchyma were delineated. Overlying organs as the lungs or the aorta have to be excluded carefully. Time activity curves were generated and compared together after background subtraction especially in case of the liver region. The background correction is necessary when colloid scintigraphy was done immediately before angiography. Than the representative background is found in the same ROI using time frames before the bolus has entered the region.

To analyse the corrected hepatic time activity curve the observer has to identify three points: (A) the time when the radioactivity enters the transplant, (B) the endpoint of the arterial inflow and (C) the beginning of the recirculation. When the activity enters the organ, the time-activity curve arises constantly. To find the endpoint of arterial inflow into the liver it is useful to measure the time necessary in case of an exclusively arterial perfused organ. Under regular conditions (normal ejection fraction of the heart) 7 ± 2 s were needed, which have to be added to time (A). Really, a notch in the time-activity curve of the liver in this time period is demonstrable at this point, when the arterial inflow ends slowly and the portal-venous inflow begins rapidly. Furthermore about 7 ± 2 s are needed to reach the beginning of the recirculation (Fig. 3).

Then the integral (I_{art}) from A to B is calculated. The count rate at B is subtracted from the hepatic curve, and the integral (I_{pv}) from B to C is calculated. The arterial fraction of total hepatic perfusion is $I_{art}/(I_{art} + I_{pv})$.

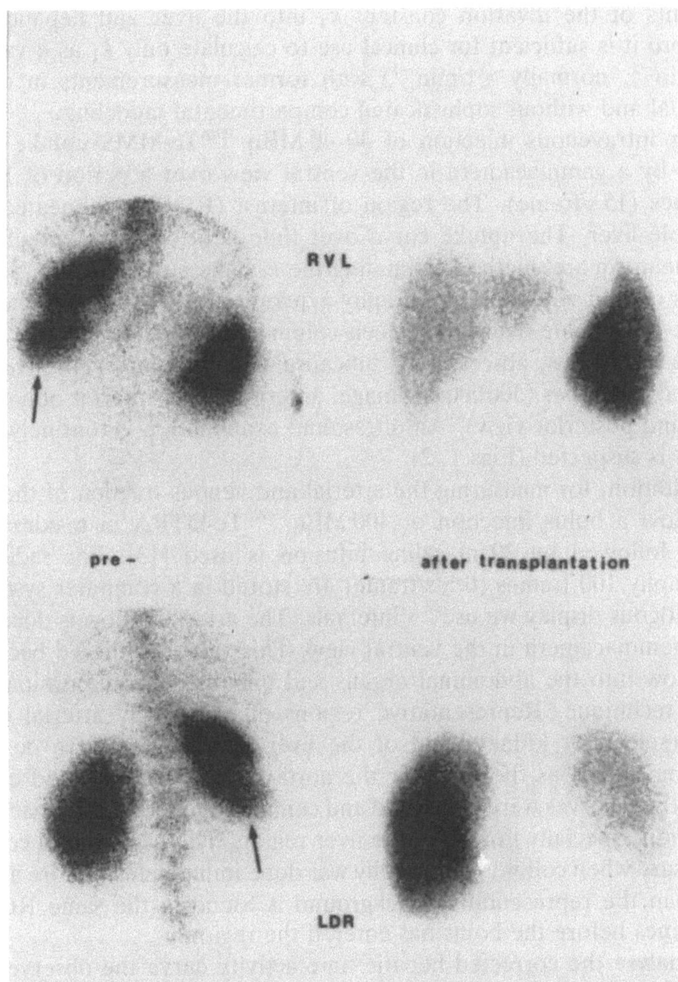

Figure 1. Liver scintigraphy before and after liver transplantation in a patient with posthepatic liver cirrhosis: Typically, the liver is very small (arrow), the splenic volume is increased and the bone marrow uptake is demonstrable (k-value < 0.4 min^{-1}, arterial fraction of the liver perfusion > 85%). After grafting the arterial fraction is decreased, but has not reached the normal level (69%). The colloid clearance (0.98 min^{-1}) and the bile kinetics (see Fig. 5) have been normalized.

Findings and problems

Normal ranges of the k-value vary from 1.2 to 1.4 min^{-1}, while in the grafted liver normal uptake is slightly diminished to about 1.0 min^{-1} in the early postoperative phase. Within one year the k-value in symptomfree patients drops automatically to 0.7 min^{-1} with a further decrease to 0.5 min^{-1} in later

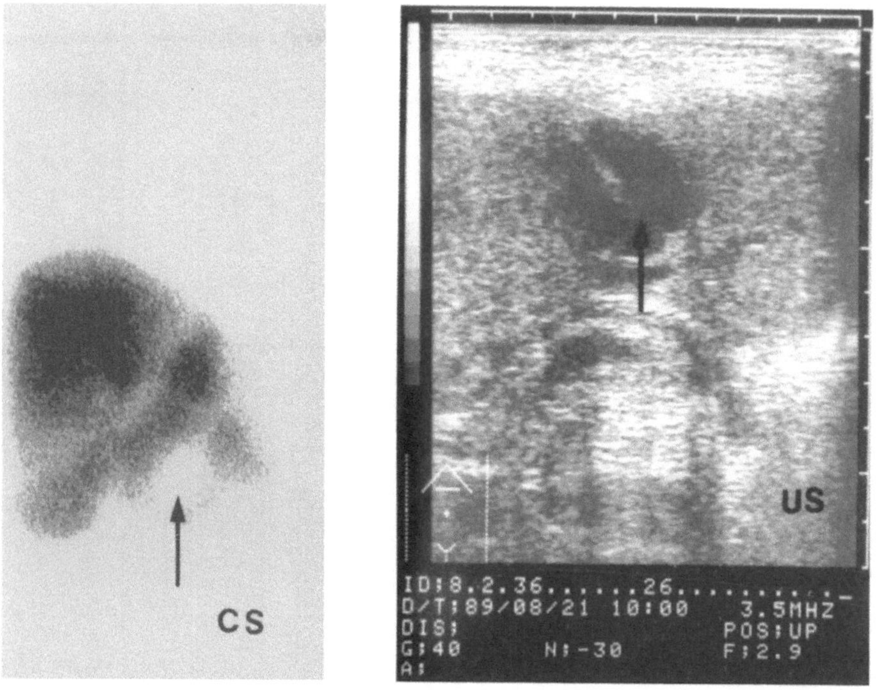

Figure 2. Demonstration of a defect in the left liver lobe (ventral view) in a patient after liver transplantation because of liver failure due to NANB-hepatitis: Ultrasonograhically, this tumor was as hypochoic as fluid and septa could be demonstrated.

years. This decrease in *k* over time may be interpreted as sign of a (minimal) chronic rejection. The sign of chronic rejection in autopsy and hepatectomy specimens is an obliterative arteritis that most extensively affects the medium-size, hilar hepatic arteries [14]. Arteriography is able to demonstrate slow hepatic arterial flow as a result of multiple, hemodynamically significant stenoses in series and consecutively increased peripheral resistance within the rejecting graft [15].

Under regular conditions the arterial fraction of hepatic blood flow is about 30% (28 ± 6%, normal range 20–40%, Martin-Comin et al. 1988) [16]. In compensated cirrhotics this fraction is increased up to 60% and comes up to 90–100% depending on the grade of decompensation.

After grafting different courses of *k*-value and arterial fraction occur due to the preoperative abdominal blood flow situation. In end-stage cirrhotics with preoperatively increased hepatic pressure and consecutively elevated arterial fraction the *k*-value is decreased to about 60% of the normal level at the first days after transplantation and normalizes within 10–14 days, if complications do not occur. In the first one to two years after transplantation the *k*-value decreases to the initial value (65–85% of normal level). The

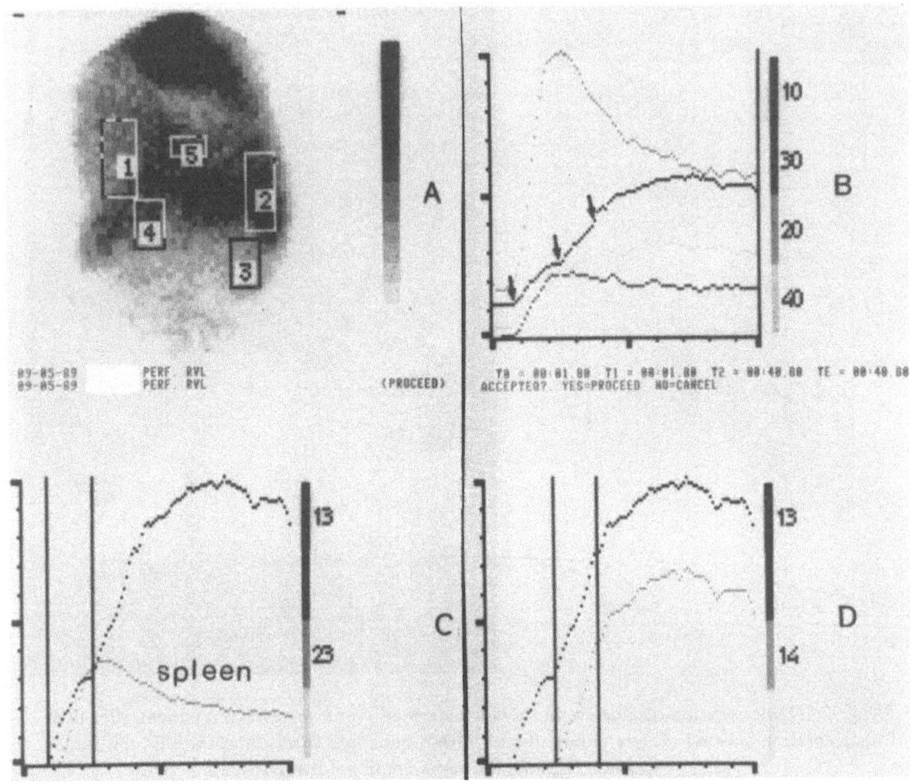

Figure 3. Estimation of the arterial fraction of liver perfusion (50%): A) Regions of interest, B) curves, C) arterial phase, D) portal-veneous phase.

arterial fraction is very high (up to 80%) and will never reach normal values (from 68 ± 8% preoperatively to 63 ± 5% at day 3–5 postoperatively to 51 ± 4% in the later course). Other authors evaluate comparable data in case of an uncomplicated transplantation (35–66%; Martin-Comin et al. 1988) [16].

In tumor patients or children with biliary atresia the abdominal blood flow situation is normal preoperatively. Thus *k*-value and arterial fraction stay normal from the first days after transplantation (*k*-value 0.8–1.2 min^{-1}, 30–40% arterial fraction) or reaches normal values within 6 to 10 days. In the early postoperative phase a graft swelling due to the extracorporal manipulations may occur leading to a diminished nutritive perfusion and need for an increased arterial flow up to 65%.

The case of an occlusion of the portal vein occurs in a frequency lesser than 1%. In the flow study a portal-venous segment of the bolus-curve is not evaluable. The time-activity course over the liver region equals those of an exclusively arterial perfused organ.

In case of a thrombosis of the hepatic artery (frequency 4%) the *k*-value decreases followed by a complete loss of colloid fixation some days later. Thrombosis of smaller intrahepatic arteries led to circumscript necrosis demonstrable as defects in the liver-spleen scan. Comparing the arterial reference curve after bolus injection to the liver curve a time lag of about 7 s occurs and a distinction of the 'arterial' and portal-venous segment is not possible.

A differentiation to an acute rejection in case of both vascular complications is not possible with this method. An emergency retransplantation has to be considered in all cases and should be done before TCT or ultrasound demonstrates focal alterations. In consequence, further diagnostics are indicated to prepare retransplantation as soon as necessary. The possibility to find a transplantable organ arises with the time having at one's disposal. A retransplantation in a stable state of the patient has a better prognosis too. The advantage of radionuclide methods is the possibility to repeat the investigation whenever indicated by clinical or laboratory course even in case of suspected hints. However, newer methods as duplex sonography seem to be established as the method of first choice to detect vascular complications. Absent hepatic artery pulse is highly diagnostic for occlusion [18].

Histologically, rejection is manifested by a cellular-mediated injury of hepatocytes and bile ductules and a spectrum of vascular lesions in the hepatic arteries [14]. The allograft is frequently swollen, caused by an inflammatory infiltrative process within the portal tracts and subsequent edema, which increases intrahepatic pressure and decreases total hepatic blood flow.

In acute rejection the arterial fraction increases analogous to the intrahepatic pressure and the *k*-value or the IDA-uptake respectively drops dramatically about 1 day before the expression of liver enzymes [19]. A controllable situation is indicated by a decrease only to 1/2 or 1/3 of the initial postoperative values. The normalization of the arterial fraction in tumor patients indicates first the success of treatment. After an acute rejection the preliminary plateau of the *k*-value is never reached, which may be interpreted as a sign of irreversible damage and loss of Kuppfer cells or its function. Prolonged (chronic rejection) or accelerated organ failure is the indication for an urgent retransplantation.

Earlier reports of postoperative care have been in the pre-cyclosporin era and have highlighted the incidence of steroid related deaths from infection, biliary complications and bleeding ulcers. Today, patients receive cyclosporin A and low dose steroids. Problems with the biliary anastomosis and acute rejection remain causes of significant morbidity but rarely of mortality. Biliary complications occur in the early postoperative phase in less than 12% (leak 8%, obstruction 4%) [20]. The Hannover unit have described improved results with the side-to-side choledocho-choledochostomy [21]. Since then the request for diagnosis a biliary complication is decreased in our department.

For demonstration a bile leakage (Fig. 4) cholescintigraphy is superior to other radiologic methods and is independent on a lying *T*-tube or intraoper-

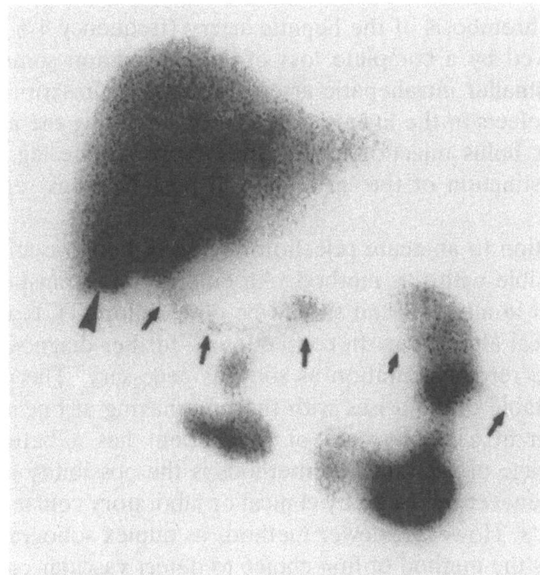

Figure 4. Cholescintigraphy for detection of bile leaks: The leakage at the hepatico-jejunostomy is still drained by a pigtail-catheter (arrow), but a bile depot is just demonstrable in the subhepatic area (arrows). The bile transport via the intestine is regular. An extension of bile into the peritoneal space could be excluded.

atively positioned catheters. In regard to 487 examinations sensitivity is calculated up to 93% and specifity up to 94% [22, 23].

The following statements could be done by observing the cholescintigraphic images:

1. Exclusion of an intra- or extrahepatic bile leakage,
2. Confirmation of a bile leakage in suspected cases (fluid filled lesion in the liver; in transplanted patients bile leakage may occur without or with minimal clinical symptoms only),
3. Hint to the source of the bile leakage and demonstration of the intra abdominal extension (intrahepatic vs. extrahepatic localized, with or without extension/diffuse extension),
4. Proof of the drainage function of lying catheters (estimation of the volume remaining intraabdominally perhaps extending locally or diffuse or transported regullary in the intestine).

Smaller leakages stopped themselves without clinical complications. Bile depots could be drained by ultrasound guided positioning of catheters. In case of a widely extension of bile and/or signs of a beginning peritonitis an urgent surgical intervention is indicated.

For exclusion of an obstruction of the biliary tract or a stenosis of the anastomosis with cholescintigraphy hepatocellular function has to be sufficient. When the parenchymal function is impaired, IDA-derivates remain in

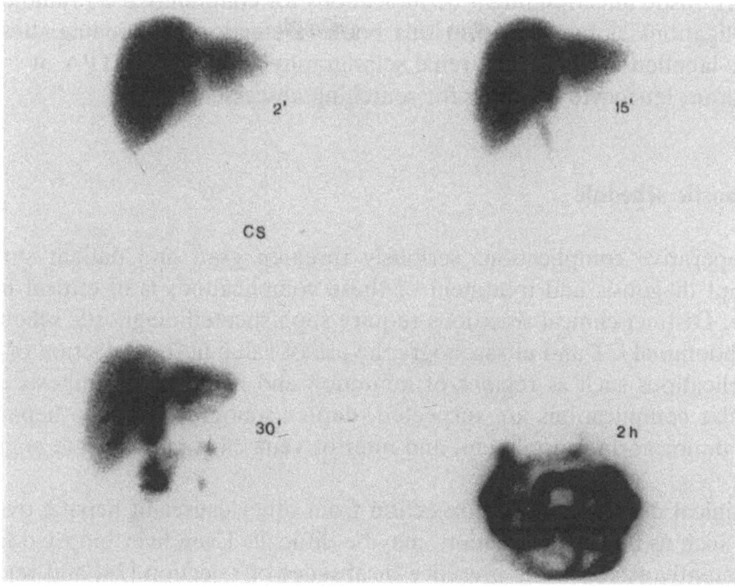

Figure 5. Cholescintigraphy for exclusion of a biliary tract problem: Images have been perfomed in the ventral view 2 min p.i. (parenchymal phase), 15 min p.i. (beginning of the bile excretion), 30 min p.i. (bile flow via the choledocho-choledochostomy is not involved) and 120 min p.i. (bile kinetics in time, exclusion of a bile leakage).

the blood pool and an increased fraction is cleared by the kidneys. In the liver the tracer remains in the blood pool too and enters the hepatocytes slowly. A small portion per time is cleared and transported via the biliary tract leading to no or minor biliary duct contrast but demonstrable intestinal activity on the later scans (24 h after application). If any tracer is not demonstrable even on later scans, a total obstruction is probable. Due to cholestasis an impairment of the hepatocytes is expected proportional to the duration of obstruction. The cholescintigraphic differentiation between both entities (hepatocellular icterus vs. obstructive icterus with consecutive hepatocellular impairment) is not clear-cut.

If the transplanted organ is able to transfer the tracer a bilio-dynamic considerable obstruction could be excluded by normal IDA-kinetics or the stop could be localized either in the intrahepatic tree or in the extrahepatic anastomosis (Fig. 5). In case of a stenosis tracer kinetics are delayed (intestinal occurence of the tracer later than 45 min after application) and the stop is demonstrable constantly as a 'tapering reservoir.' For clinical use the radionuclide methods have the advantage to exclude an obstruction without any risk for complications but the disadvantage to work only in not really impaired transplants.

Other complications after liver transplantation as bleeding, renal failure, opportunistic infection might be indications for commonly used radionuclide investigations as mentioned in this book (Detection of bleeding sites with ^{99m}Tc-labelled erythrocytes, renal scintigraphy with ^{99m}Tc-DTPA or $^{123/131}$I-hippuran, leucocyte scanning for searching abscesses).

Diagnostic schedule

Postoperative complications seriously threaten graft and patient survival. Prompt diagnosis and treatment of these complications is of critical importance. Distinct clinical situations require sophisticated diagnostic schedules.

Abdominal CT and ultrasonography are of value in the detection of many complications such as regions of infarction and necrosis, thrombosis etc. If vascular complications are suspected, duplex sonography of the hepatic artery, donor aorta, portal vein, and inferior vena cava may obviate angiography.

Clinical differentiation of rejection from other causes of hepatic dysfunction, such as biliary obstruction, may be difficult. Even liver biopsy does not consistently establish the presence or absence of rejection [24] and rejection remains a diagnosis of exclusion. In this context, studies of allograft perfusion and bile flow with radionuclides are helpful.

Cholescintigraphy is specifically used to diagnose bile leakage. Focal fluid collections as biliomas, hematomas and seromas are seen with relative frequency after liver transplantation. Directed fine needle aspiration differentiates all these cases and is able to document infection.

References

1. Dindzans VJ, Schade RR, Gavaler JS, Tarter RE, van Thiel DH (1989) 'Liver transplantation: A primer for practicing gastroenterologists, part I.' *Dig Dis Sci* 34: 2–8.
2. Starzl TE, Iwatsuki S, Shaw B, Gordon R (1985) 'Orthotopic liver transplantation in 1984.' *Transplant Proc* 17: 250.
3. Singer PA, Siegler M, Whitington PF, Lantos JD, Emond JC, Thistlethwaite JR, Broelsch CE (1989) 'Ethics of liver transplantation with living donors.' *NEJM* 321/9: 620–622.
4. Pichlmayr R, Neuhaus P (1985) 'Lebertransplantation.' *Chirurg* 56: 211.
5. Hawkins RA, Hall T, Gambhir S, Busuttil R, Huang S, Glickman S, Marciano D, Brown R, Pelps ME (1988) 'Radionuclide evaluation of liver transplants.' *Sem Nucl Med* 18: 199–212.
6. Chervu LR, Nunn AD, Loberg MD (1982) 'Radiopharmaceuticals for hepatobiliary imaging.' *Sem Nucl Med* 12: 5–17.
7. DeJonge M, Pauwels E, Hennis P (1983) 'Cholescintigraphy and Tc-99m-diethyl-IDA for the detection of rejection of auxiliary liver transplants in pigs.' *Eur J Nucl Med* 8: 485–488.
8. Taylor K, Morse S, Weltin G, Riely C, Wayne M (1986) 'Liver transplant recipients: portable duplex US with correlative angiography.' *Radiology* 159: 357–363.
9. Krishnamurthy S, Krishnamurthy GT (1989) 'Technetium-99m-Iminodiacetic acid organic anions: Review of biokinetics and clinical application in hepatology.' *Hepatology* 9: 139–153.

10. George EA, Hendershott Lr, Klos DJ, Donati RM (1980) 'Mechanism of hepatic extraction of gelantinized ^{99m}Technetium sulphur colloid.' *Eur J Nucl Med* 5: 241–245.
11. DeNardo S, Bell GB, DeNardo GL, Caretta RF, Scheibe PO, Imperato TJ, Jackson PE (1976) 'Diagnosis of cirrhosis and hepatitis by quantitative hepatic and other reticuloendothelial clearance rates.' *J Nucl Med* 17: 449–459.
12. Dobson El, Jones HB (1952) 'The behaviour of intravenously injected particulate material; its disappearance from blood stream as a measure of liver blood flow.' *Acta med Scand* 144: *Suppl.* 273, 1–7.
13. Biersack HJ (1980) 'Die quantitative Leberperfusionsszintigraphie.' *Arch Klin Chir* 305: 23.
14. Demetris AJ, Lasky S, van Thiel DH, Starzl TE, Dekker A (1985) 'Pathology of hepatic transplantation. A review of 62 adult allograft recipients immunosuppressed with cyclosporine/steroid regimen.' *Am J Pathol* 118: 151–161.
15. Morse SS, Reuben A, Strauss EB, Greenwood LH, Denny DF, August DA, Fleye MW (1986) 'Liver transplant rejection arteritis: Serial hepatic arteriography.' *Cardiovasc Intervent Radiol* 9: 191–194.
16. Martin-Comin J, Mora J, Figueras J, Puchal R, Jaurrieta E, Badosa F, Ramos M (1988) 'Calculation of portal contribution to hepatic blood flow with ^{99m}Tc-microcolloids. A noninvasive method to diagnose liver graft rejection. *J Nucl Med* 29: 1776–1780.
17. Brölsch CE, Creutzig H, Neuhaus P, Pichlmayr R (1981) 'Leberdurchblutung nach orthotoper Lebertransplantation bei Cirrhotikern und bei Tumorpatienten.' *Langenbeck's Arch Chir Suppl* 259–263.
18. Segel MC, Zaijko AB, Browen A, Bron KM, Skolnick ML, Penkrot RJ, Starzl TE (1986) 'Hepatic artery thrombosis after liver transplantation: Radiologic evaluation.' *AJR* 146: 137–141.
19. Creutzig H, Brölsch C, Müller S, Neuhaus P, Gratz KZ, Gonda S, Schober O, Schwarzrock R, Pchlmayr R, Hundeshagen H (1984a) 'Follow-up of liver transplanted patients with radionuclides,' in: Lutz H, Demling J (eds), *Diagnostic Imaging Methods in Hepatology*. pp 143–145. Lancaster: MTP Press.
20. Kirby RM, McMaster P, Clements D, Hubscher SG, Angrisani L, Sealey M, Gunson BK, Salt PJ, Buckels JAC, Adams DH, Jurewicz WAJ, Jain AB, Elias E (1987) 'Orthotopic liver transplantation: postoperative complications and their management.' *Br J Surg* 74: 3–11.
21. Neuhaus P, Brölsch C, Ringe B (1984) 'Results of biliary reconstruction after liver transplantation.' *Transplantation Proceedings* 16: 1225–1227.
22. Gratz KF, Creutzig H, Brölsch C, Pichlmayr R, Hundeshagen H (1982) 'Szintigraphie des Gallelecks.' *Radioakt Isotope Klinik Forsch* 15: 189–194.
23. Creutzig, H, Brölsch C, Müller S, Neuhaus P, Gratz K, Schober O, Pichlmayr R, Hundeshagen H (1984b) 'Nuklearmedizinische Diagnostik des Gallelecks.' *DMW* 109: 1398–1400.
24. Scharschmidt BF (1984) 'Human liver transplant: Analysis of data on 540 patients from four centers.' *Hepatology* 4: 955–1015.
25. Gambhir SS, Hawkins RA, Huang S, Hall TR, Busuttil RW, Phelps ME (1989) 'Tracer kinetic modeling approaches for the quantification of hepatic function with Tc-99m DISIDA and scintigraphy. *J Nucl Med* 30: 1507–1518.

7. Differential diagnosis of liver tumors

KLAUS F. GRATZ, OTMAR SCHOBER
and BURCKHARD RINGE

Introduction

Since ultrasound (US) has been widely used in clinical routine diagnostic of the liver and abdomen, it is easy to demonstrate a livertumor in an early and presymptomatic stage. Indications for US examination could be thorough basic diagnostic routinely done as screening or in connection with other, mostly malign diseases and abdominal pain of unknown origin. It is necessary to make an in-vivo differential diagnosis, since the different types of liver tumors have a different prognosis and therefore none, less or more therapeutic interventions are indicated enhancing the possibility and extent of complications.

Is the tumor delineated without doubt, one has to reflect (Table 1) on the question, is this tumor a liverown proliferative process or a more generalized disease with hepatic manifestation as metastasis in the liver of other origin, an abscess, dysontogenetic or traumatic cysts, Caroli syndrome, a liver infarct with bile leakage, a lipomatosis or focal liverfat. As the first step examination US gives the first hints in demonstrating non-hepatic metastasis too or in establishing the diagnosis of an liquid tumor. Biopsy and cytological examination verify the diagnosis of bile leakage (bile), cystfluid, malign or infectious material. In these cases the accuracy of liverbiopsy is about 80–90% and has a high clinical prediction in positive findings.

Liver own tumors were classified by histological criteria considering the tissue origin [1]. Hemangioma and hemangioendothelioma are of mesenchymal origin, the focal nodular hyperplasia (FNH), the hepatoblastoma, hepatocellular adenoma (HCA) or carcinoma (HCC), intrahepatic bile duct cystadenoma or carcinoma are epithelial tumors. Only the hemangioma and the FNH have an unhesitating prognosis. All other tumors should be diagnosed definitively by histological examination. This means, the tumor has to be resected if possible. To answer the question of resectability radionuclid procedures contributes little. US, transmission computed tomography (TCT), magnetic resonance imaging (MRI) and angiography are necessary in this case.

H.J. Biersack and P.H. Cox (eds), Nuclear Medicine in Gasteroenterology, 101–117

Table 1. Differential diagnosis of liver tumors

Cystic lesions	* Bile leakage * Abscess * Metastasis
Benign solid tumors	* Hemangioma * Focal nodular hyperplasia (FNH)
Malign solid tumors	* Hepatocellular adenoma/carcinoma (HCA/HCC) * Cholangiocarcinoma * Hamartoma * Hepatoblastoma * Hemangioendothelioma, -Sarcoma

Pathology and epidemiology

The most common benign tumor in the liver is the hemangioma. The frequency at autopsy is given up to 800 cases out of 100000 inhabitants (Table 2). The lesions are frequently in a subcapsular location. Usually they are solitary (multiple up to 10% of cases). Histologically, hepatic hemangioma consist of blood-filled spaces lined by a single layer of endothelium. These spaces are separated by fibrous septa, which commonly proliferate centrally and extend peripherally to a variable degree. Calcifications and thrombosis are possible. 90% of the hepatic hemangiomas are typically less than 4 cm in diameter. Tumors larger than 4 cm are classified as 'giant hemangiomas'. This may result in liver enlargement, rarely in myocardial insufficiency due to a higher shunt volume. Hemangiomas are usually asymptomatic. Solely, most of the patients complain of abdominal discomfort. Spontaneous hemorrhage has been reported in 4.5% of cases [2].

Infantile hemangioendotheliomas are made up of numerous intercom-

Table 2. Pathology and epidemiology

	Incidence	Pathology	Pathopysiology
Hemangioma	$800/10^{-5}$	90% < 4 cm Vascular network	Normal arterial flow Increased blood pool
FNH	$20/10^{-5}$	Central scar Kupffer cells Regular hepatocytes Inadequate bile ducts	Hypervascularity RES Bilirubin kinetics
HCA	$4–10/10^{-5}$	Intact capsular Atypical hepatocytes Absent bile ducts Present bile canaliculi	Proliferative Bilirubin kinetics
HCC	$7/10^{-5}$	Tendency to invade Atypical hepatocytes Variety of cytological differentiation and histological pattern	Malignancy Bilirubin kinetics Alpha fetoprotein

municating vascular channels, larger vessels and bloodfilled spaces lined by endothelial cells. This tumor is locally aggressive. Death may result from congestive heart failure, replacement of liver parenchyma or, rarely, from metastases. Epitheloid hemangioendothelioma [3] tends to infiltrate and obliterate hepatic and portal venous channels, leading to extensive hepatic fibrosis. Pleomorphic cells with convoluted hyperchromatic nuclei, in parts forming loose irregular tubular structures permeate the hepatic sinusoids. A hemangiosarcoma is composed of spindle cells with growth into the lumina of pre-existing vascular spaces such as liver sinusoids and small veins. The liver cells may eventually atrophy and disappear. In some instances large, cavernous spaces exist formed with solid, spindle cell masses.

It is reported that FNH has an incidence of 20 cases out of 100000 inhabitants. This tumor can be found at any age and sex and not only in women with long-term use of oral contraceptives. The lesion is composed of hyperplastic hepatic parenchyma subdivided into nodules by fibrous septa which may form stellate scars. This tissue contains hepatocytes but shows the same loss of normal lobular architecture as in macronodular cirrhosis. The fibrous septa contain small bile ducts. Abnormally thick-walled vessels are characteristically found (hypervascularisation). 35% of the patients present abdominal discomfort or pain combined with hepatomegaly. In less than 1% the clinical course is complicated by hemoperitoneum from rupture of the tumor.

Hepatocellular adenoma were seen at autopsy in 4 to 10 per 100000 cases. They are usually well defined and may be partly or completely encapsulated. The architecture of the tumor consists of trabecula two or three cells thick seperated by sinusoids lined by endothelium without Kupffer cells. The atypical cells are generally larger than hepatocytes with little variation in size or shape. Portal tracts and bile ducts are absent, but bile canaliculi are present and may contain bile plugs. Numerous large vessels are invariably present. Thrombosis of these vessels leads to infarction, rupture of the tumors and intraperitoneal hemorrhage consecutively. About 44% of the patients complain of acute upper abdominal pain due to infarction, necrosis, rupture and hemoperitoneum. A malign transformation cannot be fully excluded. To differentiate this entity from a highly differentiated hepatocellular carcinoma histological examination of the total tumor is necessary.

Hepatocellular carcinoma (HCC) shows a regional variety of incidence up to 7 cases per 100000 inhabitants [4]. This incidence is proportional to the contamination of inhabitants with hepatitis B-virus. It is suggested that this may be an aetiologic factor. From animal experience it is known that aflatoxine, nitrosamine, pyrolizidin-alkaloides, butter yellow and thorotrast contrast media stimulated carcinogenesis.

In the western countries HCC is mostly associated with liver cirrhosis of many years. 75% of this type of HCC is nodular and frequently multilocal. HCC is defined as a malignant tumor composed of cells resembling hepatocytes, but a variety of histological patterns and cytological features is observed. Distinction between well, poorly or anaplastic differentiated tumors may be of biological and clinical significance. The resemblance of the tumor cells to hepatocytes varies with the degree of differentiation. Spindle-shaped cells are a sign of dedifferentiation. Bile production is seen in a minority of HCC and is specific for this tumor. Canaliculi are possibly recognizable and may be dilated. Frequently, hypervascularization is demonstrable angiographically. Metastatic spread occurs regional in the liver itself or in the abdominal lymphnodes and thereafter hematogenic towards the lungs. The prognosis is deleterious. Survival-times below one year (6 months) have been reported.

In case of HCC, alpha fetoprotein is the typical tumormarker. A sensitivity of 70 to 80% is reported. Alpha fetoprotein is not specific, because it is increased in pregnancy, hepatitis, cirrhosis and in other tumors too. The value of alpha fetoprotein is correlated to the actual proliferation rate of the tumor. An exception is the fibrolamellar carcinoma, in which less than 10% of the cases alpha fetoprotein is elevated. This entity occurs in 40% of the patients with HCC being not elder than 35 years and is mostly resectable (50–75% of the cases). Therefore better survival times have been reported (32–68 months). This tumor type resembles FNH in his architecture, which results in difficulties to differentiate these tumor entities. A central scar and areas of increased echogenicity and lower density within the neoplasm is seen in US. TCT shows hypodensity in both. Angiography demonstrated a highly vascularized tumor whose arterial phase is characterized by abnormal vessels arranged in a septate fashion. Arteriovenous shunting to the portal vein could be diagnostic.

The hepatoblastoma is a malignant tumor composed of cells resembling primitive hepatic parenchymal cells, with or without mesenchymal elements. The cells are generally organized in orderly trabeculae with canaliculi and sinusoids. Extramedullary hemopoiesis is usually seen.

Hamartomas are benign lesions composed either of loose connective tissue and epithelial elements or of a collection of bile ducts set in a fibrous stroma. Both tumors occur in childhood. 80% of the hamartomas were diagnosed within the first year of life.

Cholangiocarcinomas are derived from cells of the biliary epithelium. This malignant tumor imposes clinically by occlusion of bilary ducts and icterus.

Rationale diagnostic approach

When a liver tumor is suspected due to anamnestic data and clinical examination, in-vitro diagnostic (tumor marker) and US are indicated at first. Sonographically, extrahepatic, abdominal spread is looked for, the primary intrahepatic tumor is verified and delineated, echogenicity and structure of the tumor is examined. Cystic characteristics are a discrete, echo-free tumor with increased echo-response at the posterior wall interface. All other circumscribed lesions appeared to be solid. A decisive diagnosis cannot be derived from the ultrasonic feature. Occasionally, further ultrasonic classification is possible. A homogeneous, hyperechoic mass with well-defined margins and posterior acoustical enhancement is typical for an hemangioma. But approximately 70% of hemangiomas are truly hyperechoic. Of these about 60% are homogeneous, so that the ultrasound results may be only directive [5].

Ultrasound or TCT guided fine-needle puncture of focal lesions of the liver has become an established method for determination of their identity. While complications following puncture have been reported (in up to 3% of the cases), recent advances in biopsy and localization techniques have diminished the incidence of complications. Care should be taken when performing biopsies of lesions that are obviously hypervascular. The frequency of hemorrhage can be diminished by diagnosing first with a combination of noninvasive imaging modalities as bolus-TCT, Gadolinium-DTPA enhanced MRI or radionuclide studies. Occasionally, percutaneous biopsy will not be necessary, especially in case of benign tumors as hemangioma and FNH. In these entities any rate of lethality (or morbidity) is not acceptable.

The material obtained by biopsy is examined for cytomorphology, allowing the presence or absence of tumor cells to be determined. When no tumor cells are detected in the aspirate, it often remains uncertain whether the focus had been correctly biopsied or whether the tumor is benign and build up only by normal cells. Fresh blood is generally aspirated in case of hemangioma, which is not specific for this entity. Only if some benign endothelial cells are detected too in the cytologic smears, this fact permits a specific diagnosis. It remains to state that the cytologic evaluation of fine-needle aspirates is highly diagnostic in case of malign masses which need resection and histological proof, but is not enough predictive in case of benign tumors which need no further therapy or histological confirmation

Radiologic procedures

Radiologic methods like angiography, TCT, MRI (Table 3) have a better spatial resolution than radionuclide methods. In case of hemangioma all three methods are highly accurate by in vivo characterising this entity. A homogeneous, well-circumscribed, low-density lesion on unenhanced scans and subsequent centripetal advancing border of enhancement with the central area of low-density becoming progressively smaller, is generally recognized as being pathognomonic of hemangioma by TCT. This centripetal enhancement is believed to be due to the progressive flow of the contrast medium into the small sinusoids of the tumor. The same effect is demonstrable by MRI with Gadolinium-DTPA. In the native T2–scan hemangiomas are very signal-intense, which leads to an accurate delineation of these tumors. Nevertheless, small tumors (less than 1 cm in diameter) are problematic to characterize correctly because of respiratory artifacts. Angiography is specific but invasive and therefore restricted to use.

Table 3. Contribution of different diagnostic methods

	Advantages	Problems
US	Availability	Observer dependence
	High resolution	No proof of dignity
Cytology	Evidence of malignancy	Complications possible
		Not diagnostic in benignity
T-CT	Anatomical resolution	Minor specificity
BPS/CS	Evidence of benignity	Minor resolution
		Not diagnostic in malignity
MRI	High resolution	Costs and minor availability
Angiography	Vascularisation	Invasive/complications
Biopsy (OP)	Histological proof	Invasive/complications
RIS	Specificity in malignancy	Minor contrast and resolution
PET	Functional analysis	Costs and minor availability

Radiation burden is the only theoretical risk of radionuclide methods. Pathological functions of the tumors are measurable in-vivo by distinct procedures. Therefore this kind of methods are particulary useful to indicate benign dignity. In a diagnostic work up, radionuclide methods range between US and invasive procedures.

Blood pool scintigraphy (BPS) is highly diagnostic for hemangioma. Three phase cholescintigraphy (CS) is specific for FNH. When complications are excluded and benign dignity is proven in-vivo without doubt, both tumors needs no further therapy, especially no resection for histological proof [6, 7, 8]. Since both tumors are hypervascularized and cytological examination may be not specific, biopsy is not very helpful. US is helpful to indicate, which kind of radionuclide procedure should be used at first. BPS should be employed at first, if the tumor shows the typical US pattern of a hemangioma [9]. If other tumors are expected after US examination, cholescintigraphy may be of greatest value. This tracer is specific for hepatocellular tumors and its metastasis as well as bile-duct stenosis is demonstrable in case of cholangiocellular carcinoma or other malign liver tumors. Imino-diacetic-(IDA)-tracer-uptake or bile duct obstruction exclude hemangioma.

Radionuclide methods

Identification of tumor-cells via antigeneity is used to detect metastasis. Monoclonal antibodies directed against carcinoembryonic antigen (CEA) or CA 19–9 in case of gastrointestinal origin, or OC 125 in case of ovarian carcinoma have been introduced. The clinical value of these methods for detection of space-occupying processes in the sonographically good visible and bioptically good menageable lesions of the liver area is still limited

It is reported, that the radioimmunoscintigraphy (RIS) with ^{123}iodine-anti-alpha fetoprotein to detect HCC is of low success [10]. The authors yield a sensitivity of 52% and a specificity of 66%.

^{131}Iodine- or ^{90}Yttrium-antiferritin specifically localize in clinical and experimental hepatocellular cancer. The usefulness for therapy has been proved in a phase I study by Order and colleagues [11]. Six patients have been treated, only two have had partial remissions of their primary tumors, one of them had a complete remission of a pulmonary metastasis. Two tumors did not target anyway. In the first series of 105 treated hepatomas with ^{131}I-antiferritin the same group reported a 48% remission rate and a targeting of all tumors. The HCC may have a special neovasculature and a tumor synthesis and secretion of ferritin resulting in a selective antiferritin tumor targeting over normal tissue. Up to now common use of this method is not recommended.

Unspecific partial functions of the tumor tissue like perfusion, blood pool density, glucose utilisation, protein synthesis, gallium-uptake etc. have been proposed in multiparameter studies to monitor tumor response to therapy [12–14] or to characterize specifically the tumor entity. Local and periodical heterogenity of tumors render more difficult to differentiate significantly only with unspecific methods. In homogeneous, well differentiated tumors like hemangiomas the degree of a partial, unspecific function works pathognomically. In lesser differentiated tumors, a combination of unspecific and specific parameters is successful. For example in case of hepatocellular tumors, specific tracer uptake signalize hepatocellularity, tracer kinetics represent the degree of hepatocellular function and vascularity stated by perfusion scintigraphy is a typical condition for one entity to consider.

Labeling of red blood cells with ^{99m}Tc provides a suitable agent for blood pool scintigraphy (BPS) [15–16]. It is possible to use an in vivo-, an in vitro- or a combined in vivo/in vitro technique [17]. The in vivo-labeling with Sn-pyrophosphate pretreatment is sufficient for clinical routine diagnostics and is easy to do. 30 minutes after injection of Sn-pyrophosphate intraveneously 400 MBq ^{99m}Tc-pertechnetat is injected as bolus. The arterial inflow is registrated in 0.5 s intervals by a gamma camera linked to a computer system. Two second analog images are obtained. Radionuclide angiography is preferred to perform in anterior view, because the blood-filled abdominal organs and vessels are best separated in this projection. Only if the tumor is found in the dorsal segments of the liver, a posterior view is superior. The flow study is followed immediately by early blood pool scintigraphy taken in anterior, right anterior oblique, right lateral and posterior view. That view demonstrating best the tumor is used for sequential images every 5 min up to 25 mins after labeling. 500000 counts are collected for each view. Late blood pool images are taken 2 hours after injection using the same protocol. If the tumor could not be delineated on these scans, single photon emission computed tomography (SPECT) images are performed increasing sensitivity. A rotating gamma camera is used for SPECT image acquisition (360° rotation, 64 view angles, 30 s per projection). Transverse axial tomograms are reconstructed by filtered backprojection (ramp filter, no prefilter) and reorganized into sagittal and coronal sections (6 mm nominal slice thickness). Two slices (pixels) are coupled for 1 image.

Only one tumor (FNH) has been described containing Kupffer cells. Therefore the demonstration of phagocytosis in tumors is pathognomic for FNH [18]. To proof this condition conventional liver-spleen imaging is done with ^{99m}Tc labeled colloid. Only if the tumor is large enough (>3 cm diameter) or peripheral outlining the liver, a lack of demonstration, a cold defect, may be significant. In lesions with smaller diameters or central liver location sensitivity of colloid imaging is too low to present the tumor. In these cases a normal scan is expected per se and the liver-spleen scan is not diagnostic.

^{99m}Tc Iminodiacetic acid (IDA) derivates have been introduced by Harvey et al. [19]. Several derivates with different kinetics are available for cholescintigraphy (CS). For proper diagnosis of liver tumors the kind of derivate is not essential. IDA derivates are extracted via the sinusoidal membrane side of the hepatocytes passing the bilirubin channel. The transport process is completed by high intravasal bilirubin levels. Without transformation the radiopharmaceutical leaves the hepatocyte at the canalicular side and is excreted via biliary route into the intestine. There is no intestinal resorption of IDA derivates. It is possible to differentiate by time follow up 4 phases of cholescintigraphy: arterial inflow into the organ (first minute), parenchymal uptake (up to 10 min), biliary excretion and intestinal transport.

400 MBq ^{99m}Tc-IDA is injected intraveneously as bolus. The arterial inflow is registrated in 0.5 s intervals by a gamma camera linked to a computer system. Two second analog images are obtained. Radionuclide angiography is prefered to be performed in anterior view, because the blood-filled abdominal organs and vessels are best separated in this projection. Only if the tumor is found in the dorsal segments of the liver, a posterior view is superior. The flow study is followed immediately by early liver images taken in anterior, right anterior oblique, right lateral and posterior view. The anterior view is used for sequential images every 5 min up to 25 min after injection. 500000 counts are collected for each view and the time is noted. Later images were performed depending on this time (physical half-time corrected). For this indication it is not necessary to document the biliary excretion phase carefully. Gallbladder activity interfers with possible tumor uptake and should be minimalized by a fatty meal supporting gall bladder contraction. One serie of each view is taken later than 90 min after injection, when normal liver parenchyma is already cleared of the tracer.

Findings and problems

Between 1981 to 1983, 132 patients with hepatic tumors were observed in Hannover Medical School (Table 4). Since then about 150 patients every year were admistered for determininig liver tumor entity.

Cysts and metastasis (Table 5) were presented as cold defects by colloid liver-spleen scan or cholescintigraphy. Blood pool scintigraphy often fail to delineate the tumor, especially in cases when the mass is too small or the tissue blood content equals them of surrounding liver parenchyma.

Table 4. Patients referred for differential diagnosis of liver tumors 1981 to 1984: $n = 159$

*	No definitive diagnosis	$n = 27$
*	Cystic lesions	$n = 10$
*	Metastasis	$n = 27$
*	Hemangioma	$n = 24$
*	Focal nodular hypoplasma (FNH)	$n = 24$
*	Hepatocellular adenoma/carcinoma (HCA/HCC)	$n = 28$
*	Cholangiocarcinoma	$n = 4$
*	Miscellaneous tumors	$n = 9$
*	Liver cirrhosis	$n = 6$

Table 5. Scintigraphic findings in liver tumors

	Perfusion	Cholescintigraphy		Blood pool scintigraphy	
		Uptake	Trapping	Filling-in	Blood pool
Hemangioma	Normal	None	None	Possible	Positive
FNH	Hyper	Normal	Positive	None	Normal
HCC/HCA	Miscell.	Decreased	Possible	None	Miscell.
Metastasis	Miscell.	None	None	None	Normal

If a hemangioma is suspected, blood pool scintigraphy is the procedure of choice to prove this question (Table 5). Infrequently, an accelerated arterial inflow of the tracer is reported [20, 21]. Typically a decreased initial flow occurs, followed by progressive accumulation of labeled erythrocytes on delayed images. In hemangiomas flow is ineffectual resulting in hypoperfusion, and the red blood cells slow to flow into the mass are even slower in flowing out. In larger tumors without thrombosis the labeled erythrocytes invades the tumor from the periphery to the central region. This 'filling-in' could be demonstrated by images performed in the first 20 min after labeling (Fig. 1). The sign is highly specific for hemangiomas.

The pathophysiological possibility of increased blood pool density in tumors exists not only in hemangiomas. The case of metastasis from a colon carcinoma positively contrasted by red blood cell labeled tracer is documented in our collective [22]. Hepatoma is another lesion to demonstrate uptake greater than adjacent liver [21]. Depending on the prevalence of these tumors a specificity lower than 100% for detection of hemangiomas by positive contrasted blood pool scintigraphy should be expected. In our collective an over-all specificity of 85% has been evaluated.

The value of sensitivity depends on the volume of tumors administered and the technique used for scintigraphy. Using SPECT large lesions with a diameter more than 2 cm were detected in 96% compared to other diagnostic procedures. If tumors with a diameter less than 2 cm are examined and if only a planar imaging is used, sensitivity decreases to 50%. Using SPECT device in the same patients, sensitivity increases up to 65% (Fig. 2).

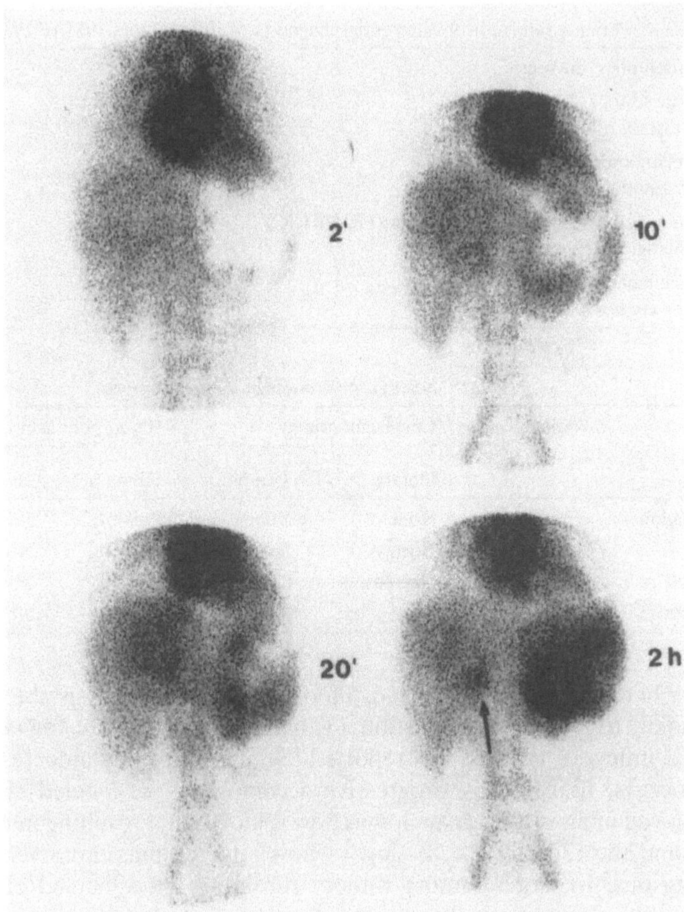

Figure 1. Blood pool scintigraphy of hemangioma: Typical filling-in of the labelled erythrocytes from the periphery of the tumor towards the center. A second tumor in the middle part of the left lobe (arrow) is only seen on the late images.

Alternative imaging methods (TCT, MRI) are not restricted by tumor diameters between 1 and 2 cm [23]. Due to respiratory artefacts smaller lesions became difficult to diagnose correctly. It is possible to prove echogenity even in very small tumors (0.3 cm). However, US is not only restricted by the spatial resolution of the instrument (0.1–0.3 cm) and transducer equipment. Time for examination and qualification of the observer play an important role in detecting and qualifying tumors.

Since specificity of BPS is not 100% using the criterion 'enhanced red blood cell uptake,' it is not recommendable to introduce this method alone for diagnosis of hemangioma. More criteria are necessary. Safe parameters are: the characteristic filling-in phenomena demonstrated by BPS or bolus-

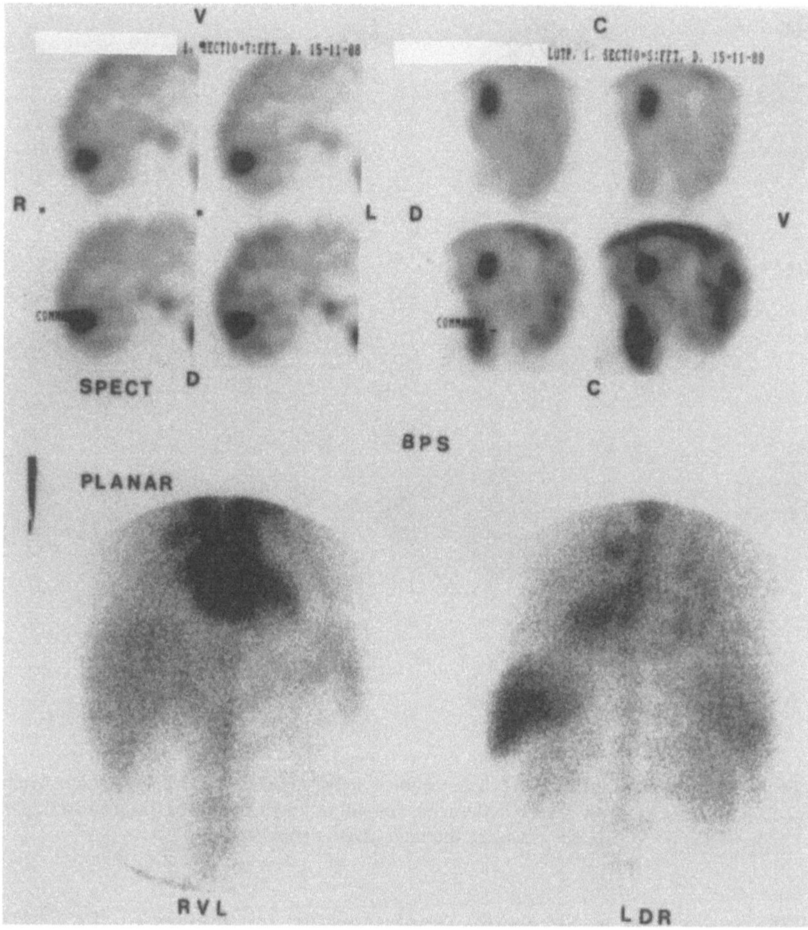

Figure 2. Blood pool scintigraphy of hemangioma: The tumor is well demonstrated in the SPECT-images, but is not seen on the planar images (ventral or dorsal view).

TCT, a typical signal intensity in T2–weighted MRI examination, benign endothelial cells in the cytologic smears in combination with a typical US feature and blood aspiration. For clinical routine it is sufficient to prove two criteria. If there is any doubt, control examinations should be performed or a resection for histological proof is indicated. An observers philosophy being on the safe side is preferable. If the criteria for a benign tumor are not clearly fullfilled, the mistake of a false-negative diagnosis could be accepted without disadvantages, but a false-positive statement (e.g.hemangioma in case of a malignoma) should be avoided.

For confirmation of FNH (Table 5) cholescintigraphy has been introduced as four phases study [24]. The diagnosis is safe, if three criteria are fulfilled:

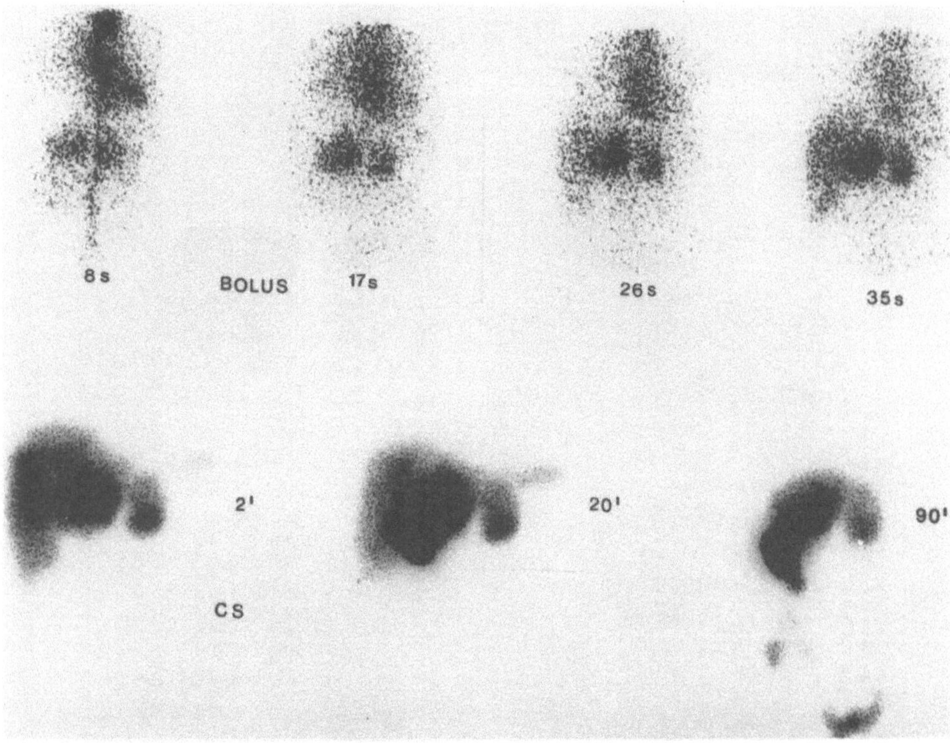

Figure 3. Cholescintigraphy of FNH: The tumor is hypervascularisized (bolus study), demonstrated a homogeneous IDA-uptake in the same amount or better than liver tissue and a trapping of the tracer in the later phase of the study.

(1) the tumor has to be positive contrasted on the images of the arterial inflow phase, (2) tumor-uptake of IDA-derivates equals or is higher than the uptake in normal liver tissue, and (3) the tumor area is positively contrasted on later images (more than 90 min p.i.) because of trapping of the IDA tracer in the tumor tissue (Fig. 3). The pathophysiological features of FNH explain the tracer kinetics: the tumor is hypervascularized, the density of hepatocytes in the mass equal the density of hepatocytes in normal liver tissue, and the bile flow out of the tumor is insufficient because of incomplete bile ducts.

If the three criteria are fulfilled, specificity is more than 98% for the determination of FNH or a benign tumor, respectively. The sensitivity of about 87% is acceptable. It is problematic to detect small tumors with a diameter less than 2 cm on the radionuclide angiogram. Another complication is the location of the lesion. Lying in projection to organs or structures with high arterial flow (as kidneys or aorta) the tumor could not be delineated. Since the same safety philosophy is used in FNH as in hemangiomas, the examination has to be declared as non-diagnostic for a benign tumor in this case. Helpful to avoid not necessary operations may be to repeat the bolus-study in an optimal position using ^{99m}Tc-HMPAO. Hexamethyl-propylene-amine-oxime (HMPAO) has the adventage to be uptaken by FNH corresponding to the increased arterial flow [25]. This allows to perform SPECT images of the trapped tracer resulting in an increased sensitivity. SPECT technique may be helpful as well in determining IDA-trapping in small tumors, if they are not detectable by planar scintigraphy.

Hepatocellular adenoma or carcinoma (Table 5) can be hypervascularized as well. Proportional to differentiation of the tumor cells uptake of IDA-derivates has been demonstrated even in metastasis of hepatocellular carcinoma [26, 27]. Therefore it is proposed to perform routinely late thoracal scans in suspected hepatocellular carcinoma for detection of metastasis. When metastasis are not demonstrable, it is not possible to differentiate correctly between an adenoma and a carcinoma by cholescintigraphy. The differentiation to FNH is nea!ly always successful. Only one case has been reported as false-positive [28]. About 30% of these tumors are not hypervascularized. In more than 86% of the cases the lesion is large enough (>2 cm) to be detectable as defect. The IDA-uptake of the mass is decreased in comparison to normal liver tissue. In about 64% of the tumors IDA-trapping is not demonstrable at all.

Since IDA uptake in the tumor region is pathognomic for hepatocellular function, a cholescintigraphic positive contrasted tumor is always of hepatocellular origin. A non-hepatocellular tumor is excluded differential diagnostically (Fig. 4). In cases of stenosis of one branch of the ductus hepaticus (e.g., a small cholangiocellular carcinoma) (see also Fig. 4) a circumscribed parenchymal retention of the tracer may be misinterpreted as tumor trapping. The discrepancy between the region of retention and the diameter of the tumor measured by US or TCT leads to the correct diagnosis. A choledochal cyst shows a retention of the tracer as well. The ultrasonographic findings are felt to be specific in this case since the dilated bile duct is noted to enter directly into a large right upper quadrant cyst.

A cockade like retention of the tracer surrounding the lesion is difficult to interpret. This pattern may reflect either a hepatocellular tumor with central necrotic, hemorrhagic or thrombotic areas or a displacing mass without hepatocellular function interfering the tracer kinetics in the surrounding liver tissue. In both cases it is no doubt, that resection of the tumor is indicated if possible.

All other tumor entities occur infrequently. The question of resectability has to bc answered first in these cases. The diagnosis is given mostly by histology.

Strategy for differentiation of liver lesions

When a liver tumor is suspected or has to be excluded, the first step (Table 6) is the US examination because of the lowest risks and costs and best availability. Based on the sonographical results malignancy is suspected (lo-

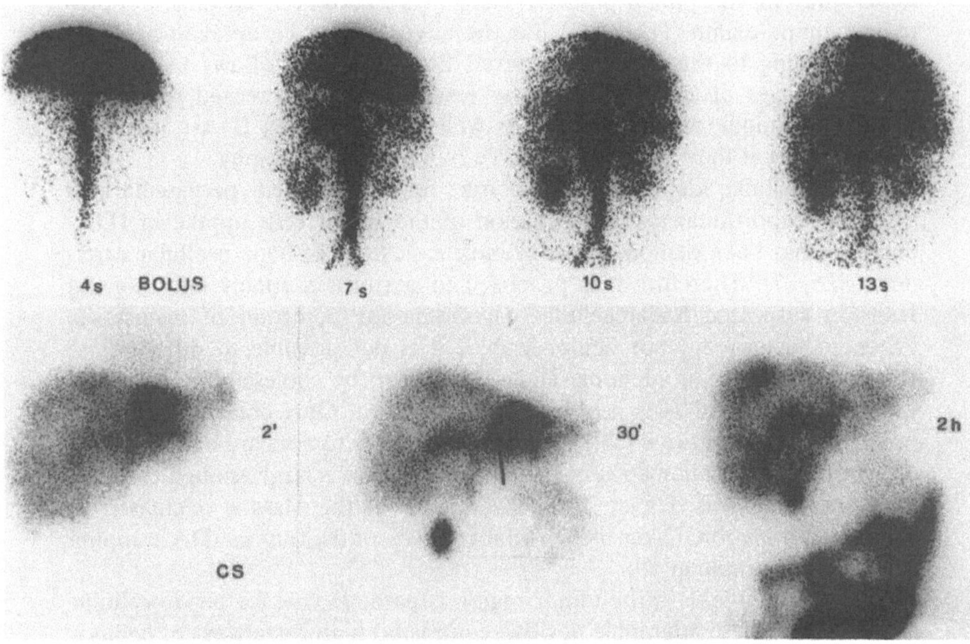

Figure 4. Cholescintigraphy of a malignant tumor (hepatoma) (US see Fig. 5D): The tumor mass invading nearby totally the right liver lobe is partially (in middle of the liver) hypervascularisized (bolus study), demonstrated an inhomogeneous, low IDA-uptake (2 min) and trapping (2 h), and causes a stop of bileflow from the left lateral liver lobe (arrow).

cation and invasiveness of the tumor, metastasis) and may be verified by biopsy, a cyst is clearly diagnosed by typical echogenity or a benign, possible hypervascularized tumor is taken into account (Fig. 5). In the last case it is decided by ultrasound whether a hemangioma or an FNH is more probable. In the case of a suspected FNH cholescintigraphy has to be performed at first. If cholescintigraphy is positive for FNH, further examinations are not necessary. If this is not the case or in case of a suspected hemangioma blood pool scintigraphy follows as next step. If an increased blood pool could be demonstrated and a second diagnostic method (US, TCT or MRI) is typical for a hemangioma, this entity is accepted. In all cases with doubt or unspecific

Table 6. Diagnostic approach in liver tumors

1.	Clinical evaluation
2.	US: Ultrasound/Cytology
3.	T–CT or BPS/CS
4.	MRI: Magnetic Resonance Imaging
5.	Angiography
6.	Laparatomy and biopsy

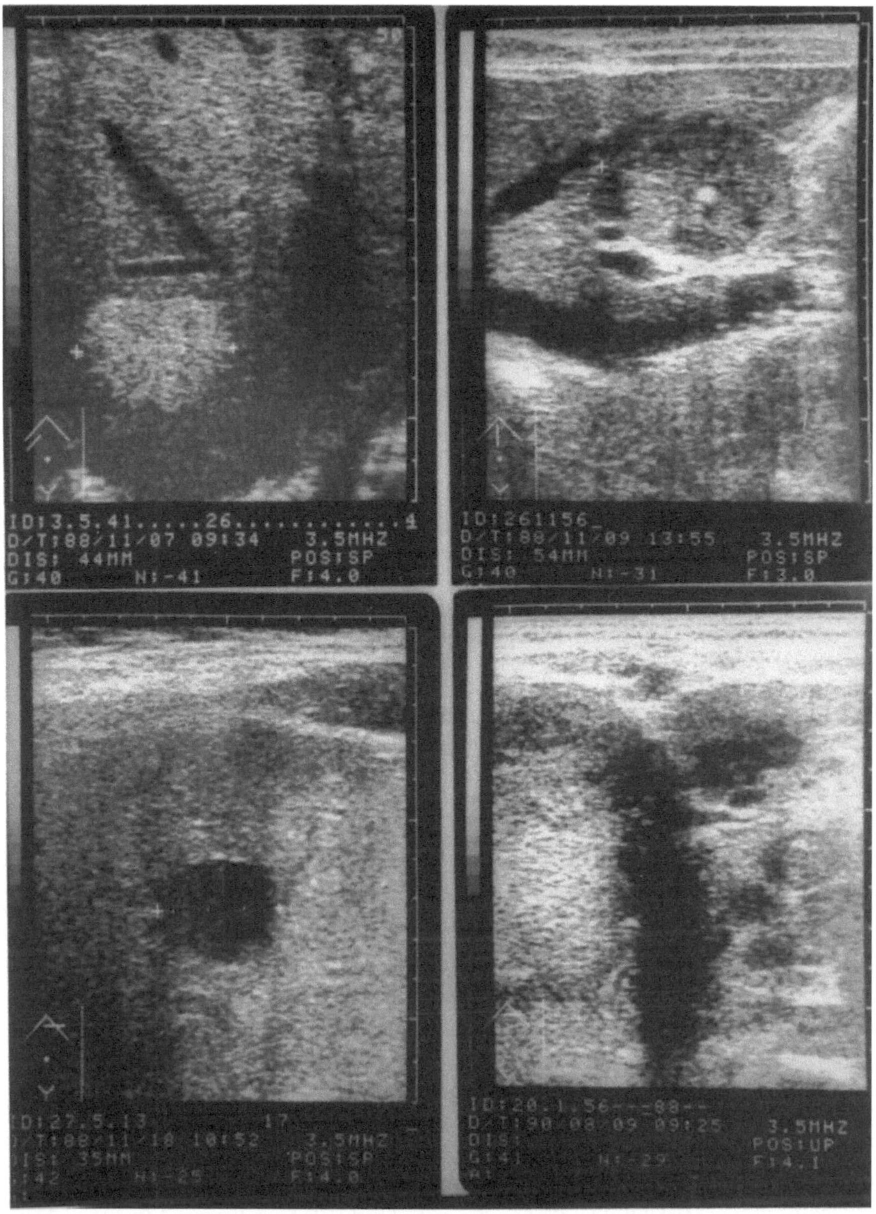

Figure 5. Ultrasonographic findings as hints for differentiation of liver tumors: (a) hemangioma (hyperechoic, homogeneous mass), (B) FNH (central scar and septa), (C) cystic lesion (hypoechoic), (D) malign tumor (left: hyperechoic, nonhomogenous) with intrahepatic metastasis (right: hypoechoic, isoechoic).

patterns difficult diagnostics including resection and histological proof are indicated.

References

1. Altmann H (1983) 'Neubildungen der Leber: Pathologie und Pathogenese.' In: Häring R (ed), *Chirurgie der Leber*, *edition medizin*, Weinheim-Deerfield, Beach-Basel.
2. Schmidt G, Börsch G, Wegner M (1985) 'Benigne Lebertumoren: Diagnostische und therapeutische Aspekte.' *Internistische Welt* 7/27: 196–200.
3. Clements D, Hubscher S, West R, Elias E, McMaster P (1986) 'Epitheloid Haemangioendothelioma. *J. Hepatology* 2: 441–449.
4. Dunham L, Bailar J (1968) 'World maps of cancer mortality rates and frequency rates.' *J Natn Cancer Inst* 42: 155.
5. Nelson RC, Chezmar JL (1990) 'Diagnostic approach to hepatic hemangiomas.' *Radiology* 176: 11–13.
6. Trastek V, van Heerden JA, Sheedy II PF, Adson MA (1983) 'Cavernous hemangiomas of the liver. Resect or observe?' *A J Surg* 145: 49–53.
7. Walt A (1977) 'Cysts and benign tumors of the liver.' *Surg Clin North Am* 57: 449.
8. Neuhaus P, Brölsch C, Ringe B, Gratz K, Majewski A, Pichlmayr R (1984) 'Diagnostik und Therapie von Lebertumoren.' *Therapiewoche* 34: 4018–4032.
9. Marchal G, Baert A, Fevery J, Klint E, Peeters S, van Dijck X, deMaeyer P, Usewils R, Wilms G (1983) 'Ultrasonography of liver hemangioma.' *Fortschr Röntgenstr* 138: 201.
10. Demangeat J-L, Manil L, Demangeat C, Rico E, Staedel-Flaig C, Duclos B, Brunot B, Jaeck D, Bellet D, Constantinesco A (1988) 'Is anti-alphafetoprotein immunoscintigraphy a promising approach for the diagnosis of hepatoma?' *Eur J Nucl Med* 14: 612–620.
11. Order S, Klein J, Leichner P, Fríncke J, Lollo C, Carlo D (1985) '^{90}Yttrium antiferritin – a new therapeutic radiolabeled antibody.' *Int J Radiation Oncology Biol Phys* 12: 277–281.
12. Hahn P, Stark D, Weissleder R, Elizondo G, Saini S, Ferrucci J (1990) 'Clinical application of superparamagnetic iron oxide to MR imaging of tissue perfusion in vascular liver tumors.' *Radiology* 174: 361–366.
13. Semmler W, Bachert-Baumann P, Gückel F, Ermark F, Schlag P, Lorenz W, van Kaick G (1990) 'Real-time follow-up of 5-fluorouracil metabolism in the liver of tumor patients by means of F-19 MR spectroscopy.' *Radiology* 174: 141–145.
14. Strauss LG, Clorius JH, Lehner B (1988) 'F-18-uracil accumulation in liver metastases after i.v. and i.a. tracer injection.' *J Nucl Med* 29 (suppl.): 776.
15. Lubin E, Zevitus Z (1972) 'Blood pool scanning in investigating hepatic mass lesions.' *Sem Nucl Med* 2: 128.
16. Engel M, Marks D, Sandler M, Shetty P (1983) 'Differentiation of focal intrahepatic lesions with ^{99m}Tc-red blood cell imaging.' *Radiology* 146: 517.
17. Front D, Israel O, Groshar D, Weiniger J (1984) '^{99m}Tc-labeled red blood cell imaging.' *Sem Nucl Med* 12/3: 226.
18. Biersack H, Thelen M, Torres J, Lackner K, Winkler C (1980) 'Focal nodular hyperplasia of the liver as established by ^{99m}Tc-sulfur colloid and HIDA scintigraphy.' *Radiology* 137: 187.
19. Harvey E, Loberg M, Cooper M (1975) 'A new radiopharmaceutical for hepatobiliary imaging.' *J Nucl Med* 16: 533.
20. Creutzig H, Gratz K, Müller S, Schober O, Brölsch C, Neuhaus, P, Lang W (1984) 'Classification of liver tumors by radionuclide imaging.' *J Nucl Med* 25: 402.
21. Rabinowitz S, McKusick K, Strauss HW, (1984) '^{99m}Tc red blood cell scintigraphy in evaluating focal liver lesions.' *AJR* 143: 63–68.
22. Gratz K, Creutzig H, Brölsch C, Neuhaus P, Müller S, Pichlmayr R, Hundeshagen H

(1984a) 'Differentialdiagnostik von Lebertumoren mit nuklearmedizinischen Verfahren.' *Radioakt Isotope Klinik Forschung* 16: 297–302.

23. Birnbaum B, Weinreb J, Megibow A, Sanger J, Lubat E, Kanamuller H, Noz M, Bosniak M (1990) 'Definitive diagnosis of hepatic hemangiomas: MR imaging versus ^{99m}Tc-labeled red blood cell SPECT.' *Radiology* 176: 95–101.
24. Gratz K, Creutzig H, Brölsch C, Neuhaus P, Majewski A, Pichlmayr R, Hundeshagen H (1984b) 'Choleszintigraphie zum Nachweis der focal-nodulären Hyperplasie (FNH) der Leber?' *Chirurg* 55: 448–451.
25. Gratz K, Schober O, Huhle T, Hundeshagen H (1988) '^{99m}Tc-HMPAO uptake by liver tumours.' *Nucl Med* 27: 117.
26. Cannon J, Long R, Berens S, Caplan G (1980) 'Uptake of ^{99m}Tc-PIPIDA in pulmonary metastases from a hepatoma.' *Clin Nucl Med* 5: 22.
27. Calvet X, Pons F, Bruix J, Bru C, Lomena F, Herranz R, Brugera M, Faus R, Rodes J (1988) '^{99m}Tc DISIDA hepatobiliary agent in diagnosis of hepatocellular carcinoma: Relationship between detectability and tumor differentiation.' *J Nucl Med* 29: 1916–1920.
28. Kotzerke J, Krischek O, Schwarzrock R, Wiese H, Ringe B, Hundeshagen H (1989) 'Focal nodular hyperplasia (FNH) and hepatocellular carcinoma (HCC) – How reliable is their differentiation by cholescintigraphy?' *Eur J Nucl Med* 15/8: 531.

8. Intra-arterial liver scintigraphy with ^{99m}Tc-MAA

RICHARD BAUER and ULRICH GEBHARDT

Introduction

Unresectable malignancies of the liver such as primary hepatocellulary carcinoma or liver metastases can be treated with systemic chemotherapy, with hyperthermic isolated perfusion of the liver, or with intra-arterial application of chemotherapeutic agents. The oldest therapeutic approach is systemic chemotherapy. Hyperthermic perfusion was proposed by Aigner [1]. However, this procedure necessitates isolation of the liver in situ and perfusion via extracorporal circulation. Short time intra-arterial perfusion with high concentration of cytostatic agents was proposed 1957 by Creech and Krementz. Soon after description of implantation of an A. hepatica catheter (1964), primary malignancies and metastases of the liver were treated by intra-arterial supplied cytostatic agents. Chemotherapeutics were administered via subcutaneous implanted arterial catheters delivering the drug into the hepatic artery [7]. Thus, the cytostatic substances perfuse the liver and their malignancies with the highest concentration, whereas the rest of the body is exposed to substantially lower concentrations. This regional chemotherapy facilitates new therapeutic concepts. It has gained increasing interest of the surgeons in oncological follow-up.

The most common source of liver metastases is colorectal cancer. Therefore, the advances of arterial hepatic versus systemic cytotoxic infusion are best investigated in this disease [11]. Preliminary results of this new therapeutic approach are looking promising [6, 12, 14, 17]. However, other investigators did not find as good results. A statistically significant improvement of selective over systemic chemotherapy with respect to reduction of the size of metastases and prolongation of median survival could not be established up to now [1, 19].

A prerequisite of optimum efficacy of intra-arterial chemotherapy are stable position of the catheter tip, permanent patency of the catheter, and maximum perfusion of the tumor [21]. To guarantee for these requirements, correct position, patency, and function of such systems have to be checked regularly as part of oncological-surgical follow-up.

Radionuclide angiography with ^{99m}Tc-labeled macro aggregated albumin

H.J. Biersack and P.H. Cox (eds), Nuclear Medicine in Gasteroenterology, 119–135

particles (^{99m}Tc-MAA) is a simple modality for demonstration of intra-arterial liver perfusion [3, 4, 13]. By means of this method, both the perfusion via the arterial hepatic catheter, the right and left hepatic arteries and their large branches, and the perfusion of the capillary bed can be assessed, where the particles are trapped during their first pass. Provided large arterial-venous shunts are absent, the amount of particle fixation is highly correlated with the perfusion of the tissue. Thus, radionuclide angiography using ^{99m}Tc-MAA allows quantitation of regional perfusion.

Radionuclide angiography (RNA)

Intra-arterial RNA of the liver using ^{99m}Tc-MAA is divided into two recording phases, dynamic first pass imaging and static acquisitions. The subcutaneous supply is localized by palpation and punctuated by means of a special needle ("Huber" type needle 22 G) preserving aseptic conditions. The needle is connected with an infusion tube. Thus, the radioactivity to be applied can be kept outside the field of view of the gamma camera.

Prior to tracer application, the hepatic catheter is rinsed with heparinized NaCI solution. If arterial blood can be withdrawn, the correct position of the catheter tip is verified. Otherwise, a test bolus of 4 MBq ^{99m}Tc-pertechnetate within a volume of 1 ml is supplied to demonstrate correct position of the catheter tip. If a position in the hepatic artery can be verified, the tracer can be administered [9].

Now, intra-arterial RNA is performed. An activity of 400 MBq ^{99m}Tc-MAA within a volume of no more than 1 ml is injected into the tube. Particle size is between 5 and 40 μm. Less than 5% of the particles are below 5 μm diameter. The tube is now a connected with a syringe with 'cold' NaCI solution. The radioactive tracer is infused with a constant flow of approximately 0.5 ml/min, which can be achieved manually with sufficient precision. This flow should be maintained during infusion, because both elevation or reduction in flow can cause artifacts. During low flow MAA particles can aggregate, whereas high flow can cause perfusion patterns different from that during chemotherapy.

First pass imaging lasts for 45 to 60 s. Data acquisition is performed with minimum 64 by 64 matrices and acquisition times of 0.5 s per frame. During blood flow through the capillaries, the macro aggregated particles are trapped. Thus, static imaging can be performed. Scintigrams are recorded in ventral, right anterior oblique (RAO45), right lateral and dorsal view. Digital resolution is 256 by 256, count rate is 2 millions per frame. If small metastases are suspected, static acquisition should be performed in SPECT technique, using a raw data matrix of 64 by 64, 60 to 64 projections divided equally over 360° with a recording time of 30 s per projection.

Thereafter, lung scans should be recorded in ventral and dorsal view. These scans reveal arterio-venous tumor shunts, if lung activity can be documented.

In cases of advanced cancer with large tumor masses, RNA shows only the tumor but does not demonstrate the remaining liver tissue. Therefore, liver scintigraphy using labeled colloids should be performed. In addition, the size of the lesion relative to the surrounding liver can be estimated. The same technique as for static MAA imaging should be used, either planar or SPECT [8, 20, 23].

Size and shape of liver tumors cannot be documented by ^{99m}Tc-MAA alone. A correct delineation of intrahepatic space occupying lesions is achieved by colloid scintigraphy. In addition, radiologic procedures such as digital subtraction angiography (DSA), angio computed tomography (ACT), or contrast enhanced nuclear magnetic resonance imaging (MRI) are applied [10, 16, 18]. DSA demonstrates the intrahepatic vessels with high resolution. However, the perfusion of the capillary bed can be assessed only indirectly by contrast agent induced changes of the lucency of the parenchyma. ACT and MRI are not applied in routine follow-up but are reserved for selected cases due to their high technical implications and high costs.

Methodology

Implantation of the arterial hepatic catheter

Angiography of the celiac trunk and of the mesenteric arteries is performed prior to surgery to document the course of the hepatic artery and their branches and reveal possible anomalies of the vessels. Correct staging with demonstration of the extension of the tumor or of the liver metastases is achieved by computed tomography (CT) or by colloid scintigraphy in tomographic technique (SPECT). Whereas normal liver tissue is mainly perfused via the portal vein and only to a small extent by the hepatic artery, perfusion of tumors and metastases is reversed. Because blood supply of liver tumors is predominated by the hepatic artery, the hepatic catheter is usually positioned into the hepatic artery [5, 15].

The catheter is fixed in the gastroduodenal artery in such a way, that its tip protrudes slightly into the lumen of the common hepatic artery. This positioning ensures only minimum interference with normal blood supply. Arterial shunts have to be definitely excluded. Therefore, arterial branches which stem from the hepatic artery like the right gastric artery or the gastroduodenal artery in which the catheter is fixed have to be ligated. Usually, the left gastric artery and the splenic artery which both originate from the celiac trunk are not perfused and can remain unchanged in situ (Fig. 1). The origin of the catheter is connected to either a port or a small infusion pump and fixed in a subcutaneous pouch.

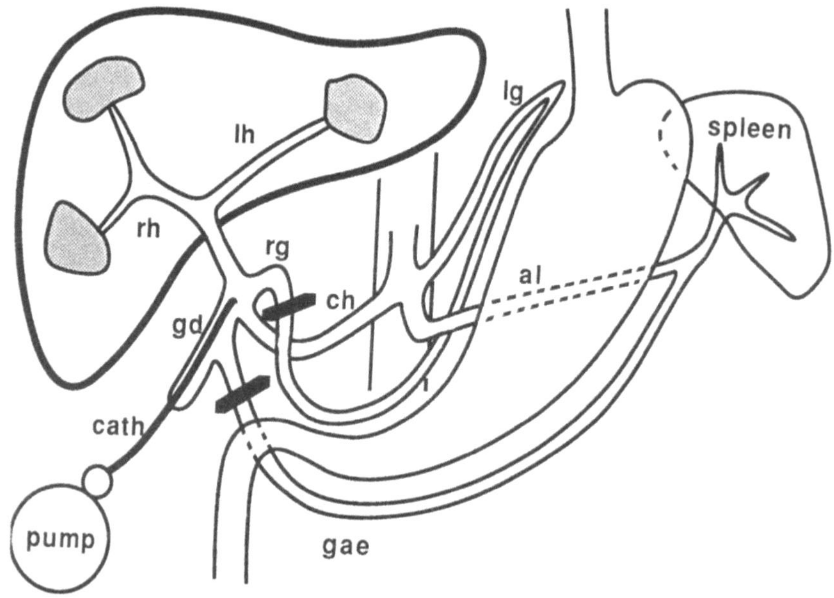

Figure 1. Scheme of selective regional liver perfusion via implanted pump and hepatic catheter (cath) delivering into the ligated gastroduodenal artery. The abbreviations mean: ch, rh, lh = common, right, and left hepatic artery; rg, lg = right and left gastric artery; gd = gastroduodenal artery; gae = A. gastroepiploica; al = A. lienalis.

Infusion systems

Two different systems can be used, an implantable pump and a port (Fig. 2). Usually, the appropriate system is selected intraoperatively. The pump, called Infusaidpump®, has a filling volume of 50 ml. An amount of 40 ml is sufficient for continuous infusion over a period of 14 days. The pressure needed is supplied by compressed gas, which expands during the 14 days lasting infusion and is compressed again during each filling with the cytostatic drugs. The Port-A-Cat® which is seen to the right of the pump, delivers into the catheter and allows for direct intra-arterial access of the liver. The small port is connected with an extracorporal infusion.

Quantification of tumor-to-lung shunts

Tumor vessels with a diameter larger than 5 to 40 μm allow for passage of the macro aggregated particles through the liver and are trapped in the capillaries of the lungs in the same way as lung perfusion scans do work. ROI's are drawn over the liver and the lungs. Activity is calculated both in ventral and dorsal view. Mean counts are determined by the square root of ventral and dorsal counts, C_v and C_d, respectively. The tumor-to-lung shunt (TLS) is obtained as

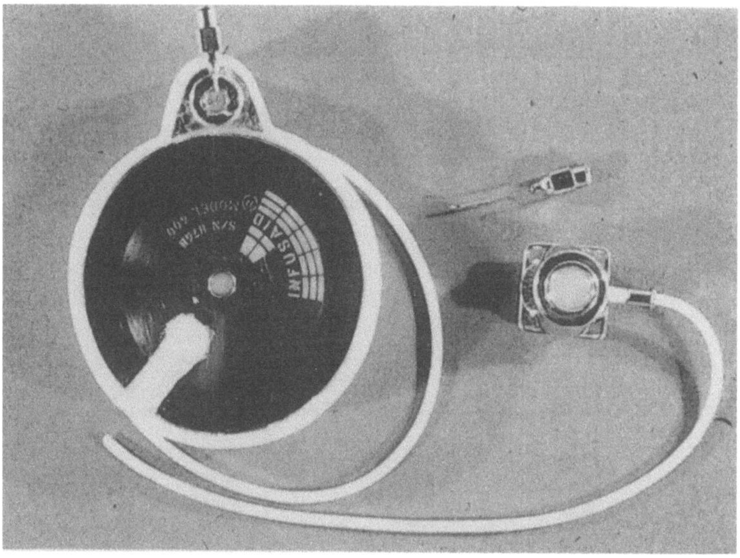

Figure 2. Implantable drug delivery systems. (Left) Infusaid® pump of 9 cm diameter, 3 cm thickness with its sideport on top of the system. The side port delivers directly into the catheter and allows for arterial drug delivery. (Right) Port-A-Cat® of 2 cm diameter. This port can be used either for short-time external infusion (prolonged injection) or for prolonged infusion by means of an external infusion pump. The Huber-type needle is seen behind the Port-A-Cat.

$$\mathrm{TLS} = (C_v \cdot C_d)^{1/2}_{\mathrm{lung}}/((C_v \cdot C_d)^{1/2}_{\mathrm{lung}} + (C_v \cdot C_d)^{1/2}_{\mathrm{liver}}) \cdot 100.$$

Quantification of the 'pathological' lung uptake can be used as an estimate of elevated systemic burden of chemotherapy. According to literature, shunt size ranges between 5 and 25% [22].

Results

Normal findings

RNA first pass

The subcutaneous supply and the course of the catheter can be assessed during the slow infusion of the tracer over a period of 45 to 60 s via the hepatic catheter (Fig. 3). The supply is demonstrated by the black dot. The inflow of the tracer into the liver reveals no anomalies. Perfusion of the liver is complete, all the particles are trapped in the liver capillaries. No substantial space occupying lesion with suspicious perfusion can be documented.

Another example of changed perfusion pattern can be seen in Fig. 4. Some loops of the implanted catheter are visualized. The catheter is correctly perfused. The caudal part of the liver shows elevated tracer uptake, which

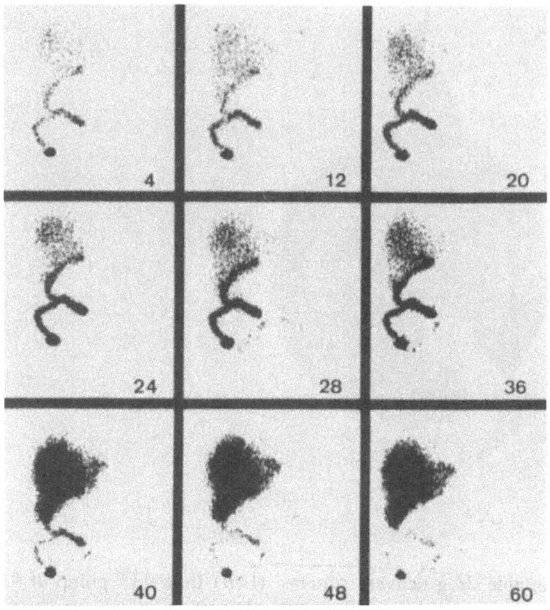

Figure 3. Radionuclide angiography (RNA) performed via a Port-A-Cat and an arterial hepatic catheter. 9 scintigrams with an acquisition time of 4 s are shown which are summed up from 8 sequential frames recorded over a period of 0.5 s each. The number in the lower right corner of each frame gives the time interval between onset of infusion and start of recording. The small black dot in the lower left corner represents the port.

is caused by a completely perfused and hypervascularized metastasis (Fig. 4). DSA reveals some more vascular details (Fig. 4). The port is to be seen on the lower left side; the tip of the catheter draining into the common hepatic artery is visualized in the middle. The right hepatic artery shows a broad lumen, the left hepatic artery is thin. DSA is usually performed for demonstration of the large vessels. The metastasis is outside the field of view, therefore, the lesion with its hypervascularization cannot be seen.

Static imaging

Scintigraphy of the trapped particles demonstrates local perfusion of the liver and its space occupying lesions. Usually, sound liver tissue shows homogeneous tracer uptake. Tumor perfusion shows three different patterns (Table 1), homogeneous hyperperfusion, hyperperfusion accompanied by a central defect and hypoperfusion. The degree of perfusion correlates with the degree of vascularization and is considered to be an important prognostic parameter for efficacy of chemotherapy [14].

Two different perfusion patterns are demonstrated in Fig. 5. The upper row 'shows typical results of a marked hypervascularized large metastasis. The colloid scan showed a large defect in the right caudal lobe of the liver

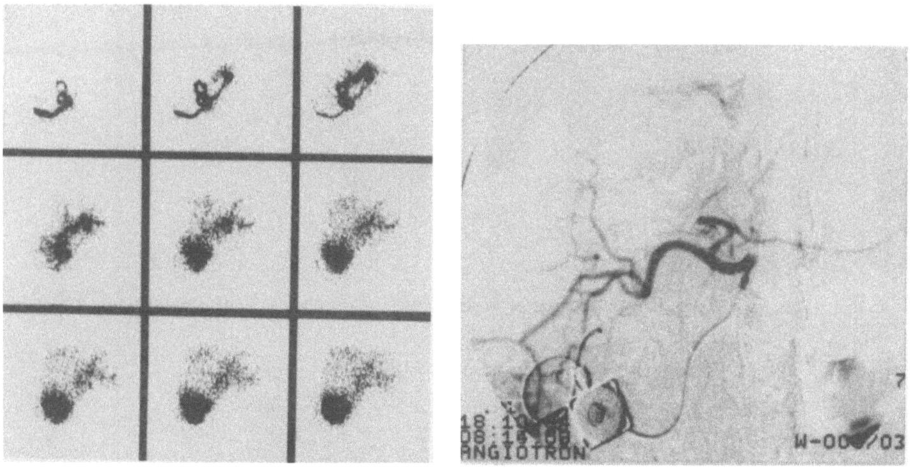

Figure 4. Hypervascularized liver metastasis. A: Nine sequential frames of RNA perfusion phase, demonstrating the delivery via two loops of the hepatic catheter, low perfusion of functioning liver parenchyma and high perfusion of a metastasis caudally in the right lobe. B: DSA shows the port in the lower left corner, the small silicon catheter draining into the common hepatic artery. The right hepatic artery has a certainly larger diameter than the left hepatic artery which branches shortly behind the catheter tip. The hypervascularized metastasis is outside the field of view.

(upper left). Thus, the normal functioning liver parenchyma of that region was replaced by tumor tissue. Perfusion with ^{99m}Tc-MAA demonstrated nearly exclusive tracer fixation within the lesion (upper right), whereas the normal liver tissue showed nearly no tracer uptake. This perfusion pattern is caused by a 'steal phenomenon'. The sound liver was perfused mainly via the V. porta, whereas the hepatic artery perfused nearly exclusively the lesion.

The lower row demonstrates findings in a hypervascularized metastasis with a central necrosis. The colloid scan revealed a large lesion in the left lobe of the liver. RNA demonstrated high perfusion of the wall of the lesion, whereas centrally no particle fixation could be seen. In addition, RNA showed two more but tiny lesions. Due to their small size, these lesions could not be found in the static, planar colloid scans.

Table 1. Typical RNA perfusion patterns as indicator of tumor vascularization (n = 65)

Perfusion pattern	Tumor vascularization	n
Homogeneously increased activity of the lesion and of the center	Hypervascularization	10
Increased activity of the periphery with central defect	Peripheral hypervascularization with necrotic center	20
Tumor periphery shows equal activity like normal liver parenchyma	Hypovascularization	35

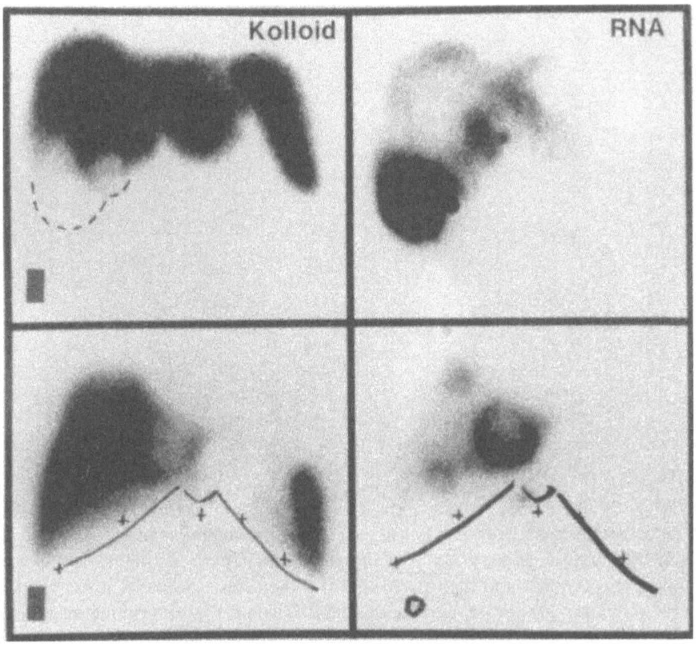

Figure 5. Planar colloid scans ('Kolloid') and RNA study in two cases with different perfusion patterns. Top: The focal defect seen in the colloid scan corresponds to a homogeneous hyperperfused metastasis. Bottom: The defect seen in the colloid scan correlates with a partially hyperperfused metastasis. However, the center of the metastasis is not perfused due to central necrosis. In addition, two further small lesions show up in the RNA study which could not be detected in the colloid scan, neither in this planar image nor in SPECT images. Note (normal) absence of the spleen in both RNA studies.

An example of hypervascularized metastases recorded in SPECT technique is demonstrated in Fig. 6. The colloid scan (left) showed multiple metastases of a medullary carcinoma of the thyroid. The contour of the liver as obtained from CT examination is superimposed to the scan. The left lobe of the liver had no tracer uptake due to marked necrosis as verified by CT. This tomogram was recorded caudally to the spleen, thus the spleen is not visualized. The RNA (right) showed complete and very high uptake of the metastases in the right lobe of the liver. According to RNA, the metastases had a high perfusion without central necrosis. The missing fixation of the labeled macro aggregated particles in the intact liver parenchyma of the right lobe is explained by a 'steal phenomenon' due to the metastases, missing perfusion of the left lobe is caused by the necrosis.

Figure 7 shows a metastasis of a hypernephroma with a hyperperfused rim and a central necrosis. In addition, the colloid scan revealed a second lesion which did not show any particle fixation in the MAA-study. This pattern was caused by a simple cyst. These last two cases demonstrate, that

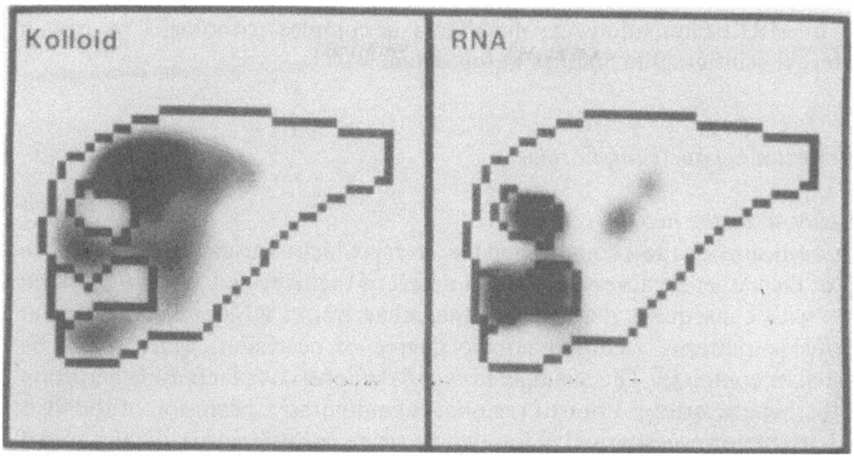

Figure 6. Two SPECT studies of liver metastases of a medullary carcinoma of the thyroid. (Left) Colloid scan shows three lesions in thr right lobe and in addition no tracer uptake in the whole left lobe. Right: RNA MAA-study shows high perfusion of the three metastases but no perfusion of the left lobe which could be suspicious for malignancy. CT confirmed necrosis of the complete left lobe.

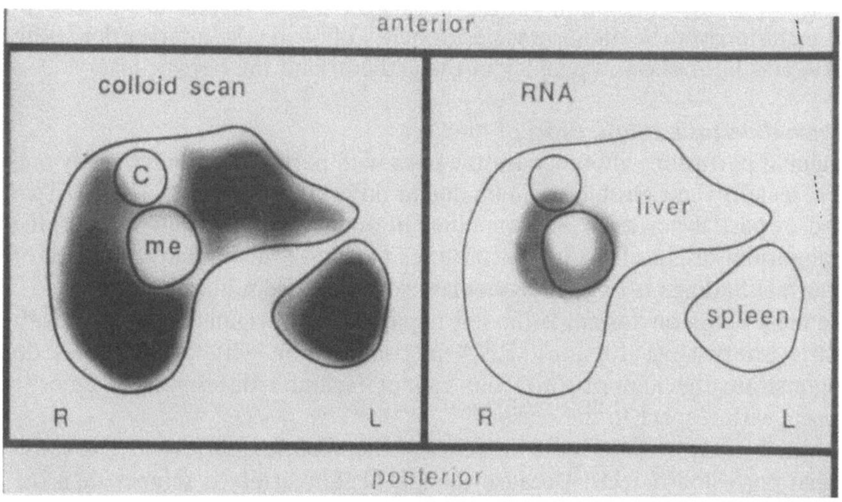

Figure 7. Transversal SPECT slices of colloid scan and RNA MAA-study of a hypervascularized metastasis of a hypernephroma ('me'). The metastasis has a hyperperfused rim but a central necrosis. An additional cyst ('C') ventrally to the metastasis produces a defect in the colloid scan but has no particle fixation of MAA. The scintigraphically assumed differentiation between these two lesions was confirmed by ultrasound and by CT.

CT or MRI examinations are mandatory in complex pathologies to explain different scintigraphic findings in full detail.

Intra-extra hepatic complications

Occlusion of the hepatic catheter

An occlusion of the common hepatic artery which is usually seen near the tip of the catheter causes a perfusion defect of the liver and reversal of blood flow with consequent perfusion of the celiac trunk. RNA reveals different perfusion patterns according to the degree of occlusion, which might be partial or complete. The example in Fig. 8 demonstrates increasing occlusion of the hepatic artery. Prior to regional chemotherapy, perfusion of the liver and its tumor was normal. However, partial occlusion was demonstrated three months after onset of chemotherapy with nearly missing perfusion of the liver, marked fixation of particles around the tip of the catheter and pathological tracer uptake by the stomach. Another four months later, no perfusion of the liver could be seen, thrombosis of the tip of the catheter had increased, and the spleen showed retrograde perfusion via the celiac trunk. Exact localization of the occlusion of the hepatic artery was demonstrated by DSA. Because no surgical correction of the catheter position could be achieved, regional therapy was discontinued and systemic cytostasis was performed instead.

Another case is given in Fig. 9. However, when RNA demonstrated occlusion of the circulation via the hepatic artery, a successful surgical re-implantation of the catheter was performed. Following the intervention, both RNA and DSA showed patency of the catheter and the hepatic artery.

Regional perfusion differences of the liver

Regional perfusion differences of the liver with partial perfusion of only one lobe, usually the right lobe, can be due to different etiologies (Fig. 10). Two third of partial perfusion abnormalities in our patient population were due to abnormal origin of the hepatic arteries (Table 2). The relative amount of abnormal findings is in good agreemant with data from literature (Table 3). The most common variant is the left hepatic artery originating from the left gastric artery (Fig. 10, top). RNA in combination with colloid scans do demonstrate the abnormalities but cannot facilitate differential diagnostic aspects with respect to the reasons.

Partial thrombosis of some branches of the hepatic arteries can also cause partial perfusion defects. Because the left hepatic artery is thinner than the right one, this vessel is more often occluded by thrombosis. Therefore, the left lobe of the liver is not perfused. Another source of perfusion defects are vascular sclerosis induced by chemotherapy or spontaneous arterial spasms.

Thus, different etiologies can cause quite similar perfusion patterns with partial perfusion defects. Differential diagnosis is impossible with scintigraphy alone. For correct scan interpretation, additional morphological infor-

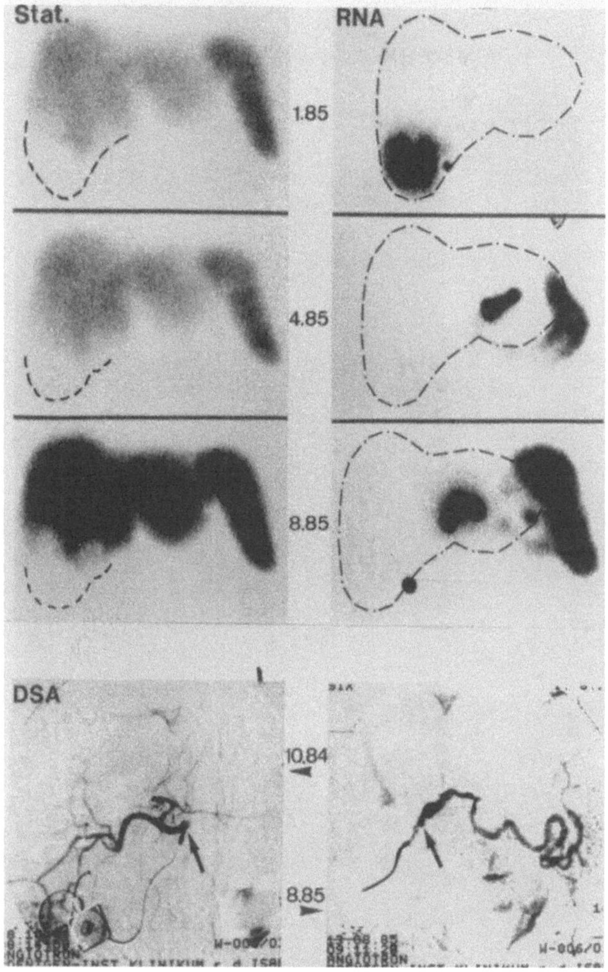

Figure 8. Demonstration of intra-extrahepatic complications during short-time follow-up. Patent catheter in January 1985 ('1.85') and hyperperfused metastasis caudally in the right lobe, corresponding to a focal defect in the colloid scan. Occlusion of the catheter three months later (4.85). Increasing thrombosis and delivery of the MAA particles to the systemic circulation, demonstrating the spleen (8.85). DSA performed in october 1984 shows patent hepatic arteries, whereas late angiography (8.85) shows complete occlusion hepatic arteries.

mation is important. These information can be obtained by DSA or by preoperative angiography of the celiac trunk and the mesenterial arteries.

Tumor-to-lung shunts

Sometimes, tumor vessels have diameters larger than the macro aggregated particles (>5 to 40 μm). Thus, MAA particles are not trapped in the capillary tumor bed but can pass this barrier, enter the venous circulation and shunt to the capillaries of the lungs were they are definitely trapped (Fig. 11).

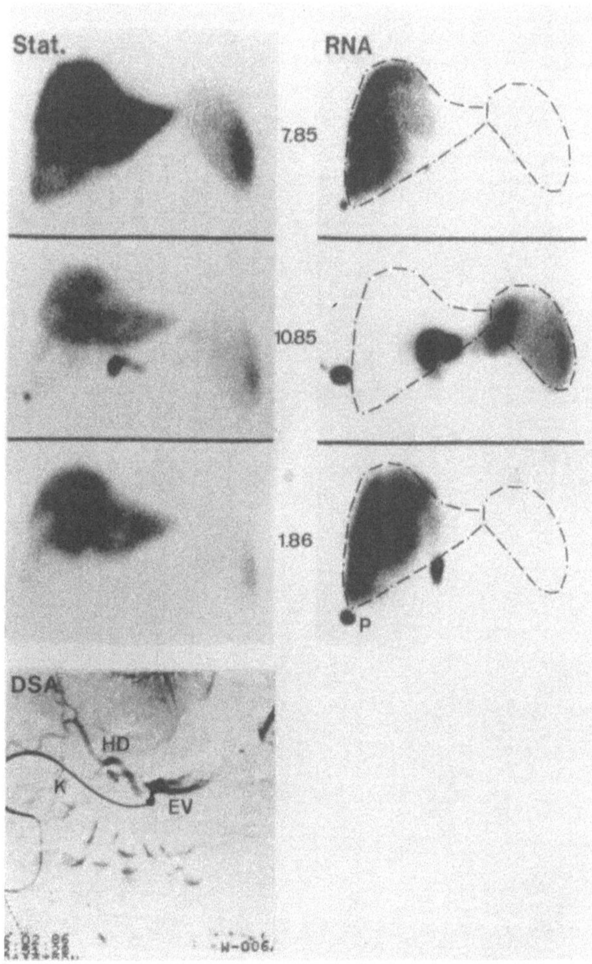

Figure 9. Successful revascularisation of an intermittent occluded catheter. In July 1985 the colloid scan ('Stat.') reveals a focal defect in the right lobe of the liver, which shows high perfusion with MAA ('RNA'). Three months later, there was no perfusion of the liver but perfusion of the spleen instead. After surgical intervention, the catheter delivers again the drug to the hepatic circulation and spleen image is absent (1.86) DSA performed some weeks later demonstrates also patent arteries.

Arterio-venous tumor shunts generate perfusion patterns of the lungs like they are seen in routine lung scans recorded in the diagnosis of lung embolism.

Tumor-to-lung shunts are often not seen at onset of intra-arterial chemotherapy but develop during treatment in 20 to 40% of the patients [3, 22]. Therefore, the lungs should be in the field of view during RNA MAA-investigations. If such a positioning is not possible, additional lung scans in dorsal and ventral view should be recorded at every RNA MAA-study.

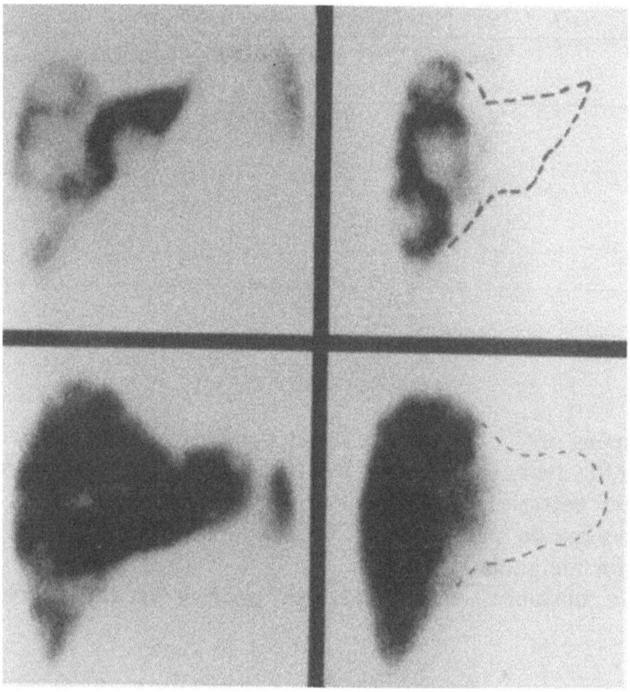

Figure 10. Partial perfusion of the liver with no perfusion of the left lobe due to two different etiologies. The colloid scans are shown on the left, RNA MAA-studies are given on the right. (Top) absent perfusion due to abnormal origin of the left hepatic artery from the left gastric artery as demonstrated by DSA. (Bottom) occlusion of the left hepatic artery by thrombosis.

Table 2. Intra- and extrahepatic complications assessed with RNA in 65 patients

Perfusion	*n*	%
Completely intrahepatic	37	57
Incompletely intrahepatic	18	27
Extrahepatic (occlusion of common hepatic artery)	5	8
Occlusion of hepatic catheter tip with extravasation	5	8
Additional arterio-venous shunts (MAA particle fixation in the lungs)	14	22

Table 3. Intra-arterial perfusion abnormalities in literature

Reports	Number of patients	Extrahepatic perfusion (%)	Arterio-venous shunts (%)
Bledin [3]	39	51	38
Ensminger [7]	13	30	–
Niederhumer [17]	106	19	(12)
Rauber [18]	44	22	–
Yang [21]	12, 17*	25	47*
Ziessmann [23]	12	–	

Discussion

RNA by means of ^{99m}Tc-MAA particles demonstrates vascularization and perfusion of the liver and its lesions by arterial blood supplied via the hepatic artery. Partial perfusion of normal liver parenchyma and of primary tumors or metastases can be quantitated. RNA can be repeated frequently without discomfort for the patient.

Cytostatic substances are infused with low flow via the arterial hepatic

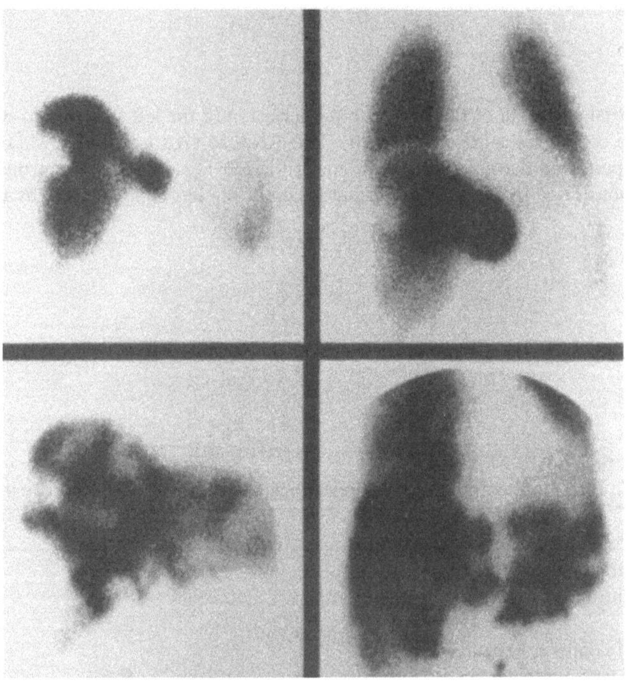

Figure 11. Demonstration of moderate (top) and marked (bottom) tumor-to-lung shunts. Colloid scans are shown on the left, RNA MAA-studies on the right.

catheter. Perfusion studies with MAA at a large scale of different flow rates have demonstrated large differences in regional partitioning of perfusion. At low flow, small vessels may be partially occluded. Therefore, the regions supplied by these vessels show no perfusion. Only at high flow these areas demonstrate good perfusion. To estimate regional distribution of the cytostatic agents between lesions and normal parenchyma, RNA investigations have to maintain the same low flow as is applied during drug delivery.

In many patients partial perfusion defects are visualized. They can be due to flow dependent differences, to thrombosis or sclerosis of vessels, or to abnormal origin of some hepatic arteries. If the perfusion of the lesions is low compared to normal parenchyma, or if the liver is not perfused at all and the cytostatic drugs enter mainly the systemic circulation, the therapeutic regimen has to be changed from high-dose selective arterial liver perfusion to normal low-dose systemic application.

The presence of tumor-to-lung shunts may also limit regional chemotherapy. In some cases, arterio-venous tumor shunts can be occluded by intra-arterial embolisation with Ivalon® particles of 200 to 250 μm size. Success of therapy can be assessed by the same radionuclide methods (RNA).

Another indication of RNA is the therapy control after temporary de-arterilization of the liver by means of rubber tourniquets [2] or short-term capillary blockage of blood supply during cytostasis by means of starch microspheres of 45 μm diameter. RNA demonstrates success or failure of interventions very easily

Conclusions

The patency of intra-arterial hepatic catheters can be demonstrated by RNA with ^{99m}Tc-MAA very easily and precisely. If partial or complete occlusion of the catheter is diagnosed, DSA should be performed for exact localization of the stop. Capillary perfusion of lesions and different perfusion patterns (hypervascularization, hypervascularization with central necrosis and hypoperfusion) can bc assessed by RNA with high precision. RNA is mandatory for revealing intra-extrahepatic complications due to occlusion or tumor-to-lung shunts. Because such complications develop frequently during therapy RNA investigations should be performed routinely.

In follow-up of tumor response to cytostatic therapy, CT and especially ACT is the 'gold standard.' However, RNA in combination with colloid scans facilitate therapy control, especially if both modalities are performed in SPECT technique.

References

1. Aigner KR, Walther H, Tonn JC, Link KH, Schoch P, Schwemmle K (1984) 'Die isolierte Leberperfusion bei fortgeschrittenen Metastasen kolorektaler Karzinome.' *Onkologie* 7: 13–21.
2. Biersack HJ, Hansen HH, Kropp J, Winkler C (1986) 'Perfusionsszintigraphie der Leber mit ^{99m}Tc-Makro-Albumin-Aggregaten (MAA) bei intraarterieller Chemotherapie von Lebermetastasen - Ergebnisse vor und nach passagerer Leberdearterilisation.' *Nuc-Compact* 17: 258–260.
3. Bledin AG, Kantarjijan HM, Wallace S, Chuang VP, Patt YZ, Haynie TP (1982) '^{99m}Tc-labeled macroaggregated albumin in intrahepatic arterial chemotherapy.' *AJR* 139: 711–715.
4. Bledin AG, Kim E, Haynie TP (1983) 'Technetium ^{99m}Tc-macroaggregated albumin angiography and perfusion.' *JAMA* 250: 941–943.
5. Bledin AG, Kim EE, Chuang VP, Wallace S, Haynie TP (1984) 'Changes of arterial blood flow pattern during infusion chemotherapy, as monitored by intra-arterially injected ^{99m}Tc macroaggregated albumin.' *British J. Radiology* 57: 197–203.
6. Daly JM, Butler J, Kemeny N, Yeh SDJ, Ridge JA, Botet J, Bading JR, DeCosse JJ, Benua RS (1985) 'Predicting tumor response in patients with colorectal hepatic metastases. *Ann Surg* 202: 384–391.
7. Ensminger W, Niederhuber J, Dakhil S, Thrall J, Wheeler JR (1981) 'Totally implanted drug delivery system for hepatic arterial chemotherapy.' *Cancer Treat Rep* 65: 393–400.
8. Gyves JW, Ziessman HA, Ensminger WD, Thrall JH, Niederhuber JE, Keynes JW, Walker S (1984) 'Definition of hepatic tumor microcirculation by single photon emission computerized tomography (SPECT).' *J Nucl Med* 25: 972–977.
9. Gebhardt U (1983) 'Angioszintigraphische Uberwachung einer selektiven Chemotherapie von Lebermetastasen.' *Der Nuklearmediziner* 10: 153–164.
10. Gebhardt U, Gmeinwieser J, Buttermann G, Heuck A, Goßmann A, Lange J, Schaff G (1989) 'Angio-SPECT und Angio CT bei intraarterieller Chemotherapie der Leber, Methodenvergleich (Abstract).' *Nucl Med* 4: 36.
11. Huberman MS (1983) 'Comparison of systemic chemotherapy with hepatic arterial infusion in metastatic colorectal carcinoma.' *Semin Oncol* 10: 238–248.
12. Hottenrott C, Lorenz M, Kirkowa-Reimann M (19??) 'Lebermetastasen: Erweiterung der Behandlungsmoglichkeiten.' *Dtsch. Arzteblatt* 36: 2539–2546
13. Kaplan WD, D'Orsi CJ, Ensminger WD, Smith EH, Levin DC (1978) 'Intraarterial radionuclide infusion: A new technique to assess chemotherapy perfusion patterns.' *Cancer Treat Rep* 62: 699–703.
14. Kaplan WD, Ensminger WD, Come SE, Smith EH, D'Orsi CJ, Levin DC, Takvorian RW, Steele JG (1980) 'Radionuclide angiography to predict patient response to hepatic artery chemotherapy.' *Cancer Treat Rep* 64: 1222–1227.
15. Lien MW, Ackermann NB (1970) 'The blood supply of experimental liver metastases. II. A microcirculatory study of the normal and tumor vessels of the liver with the use of perfused silicone rubber.' *Surgery* 68: 334 ff.
16. Lehner K, Reiser M, Gebhardt U, Heuck A, Schaff J (1987) 'DSA-Control of implanted devices for arterial hepatic perfusion.' *Cardiovasc Intervent Radiol* 10: 71–74.
17. Niederhuber JE, Ensminger W, Gyves J, Thrall J, Walker S, Cozzi E (1984) 'Regional chemotherapy of colorectal cancer metastatic to the liver'. *Cancer* 53: 1336–1343.
18. Rauber K, Lorenz M, Kirkowa-Reimann M, Hottenrott C, Riemann H (1987) 'Digitale Subtraktionsangiographie zur Kontrolle subcutan implantierbarer Katheter zur regionalen Chemotherapie von isolierten Lebermetastasen.' *Tumor Diagnostik & Therapie* 8: 11–15.
19. Rothemund M, Bruckner R, Keller E, Quint B, Knuth A, Schicketanz KH (1986) 'Regionale Chemotherapie bei Lebermetastasen kolorektaler Karzinome mit implantierbaren Gasdruckpumpen.' *Dtsch. med. Wschr.* 652–658.
20. Strauss L, Bostel F, Clorius JH, Rapteau E, Wellmann H, Georgi P (1982) 'Single-photon emission computed tomography (SPECT) for assessment of hepatic lesions.' *J Nucl Med* 23: 1059–1065.

21. Yang JP, Thrall JH, Ensminger WD, Niederhuber JE, Gyves JW, Tuscan M, K. Doan K, Cozzi E (1982) 'Perfusion scintigraphy (^{99m}Tc MAA) during surgery for placement of chemotherapy catheter in hepatic artery.' *J Nucl Med* 23: 1066–1069.
22. Ziessman HA, Thrall JH, Gyves JW, Ensminger WD, Niederhuber JE, Tuscan M, Walker S (1983) 'Quantitative hepatic arterial perfusion scintigraphy and starch microspheres in cancer chemotherapy.' *J Nucl Med* 24: 871–875.
23. Ziessman HA, Wahl RL, Juni JE, Gyves JW, Ensminger WD, Thrall JH, Keyes JW, Walker SC (1985) 'The utility of SPECT for ^{99m}Tc-MAA hepatic arterial perfusion scintigraphy.' *AJR* 145: 747–751.

PART TWO

Stomach and Intestines

9. Detection of gastroduodenal ulcers using Technetium-99m-labeled sucralfate

NICOLE A.M. PUTTEMANS, PIERRE P. ANDRE,
SERGE A.M.J. JAMSIN, DANIEL P.H. BALIKDJIAN,
and FRANÇOIS LUSTMAN

Abstract

Peptic ulcers are very common and are usually diagnosed by radiography or endoscopy. Sucralfate is an effective medication for the treatment of gastroduodenal ulcers, and Technetium 99m (^{99m}Tc) labeled sucralfate has been found to adhere to ulcers.

We examined the reliability of confirming the diagnosis of gastroduodenal ulcers, previously detected by endoscopy and histology, by means of sucralfate labeled with ^{99m}Tc, imaging one and two hours after oral administration. Fifty-four out of 64 patients were positive by scintigraphy with our method (sensitivity: 84.3%). Scintigraphy was positive in all the patients who hemorrhaged during the procedure. Our method was specific in 75% for the detection of peptic ulcers, and there were no false positive results.

The procedure was well tolerated and non-invasive. It was performed at the bedside in critically ill patients and in follow-up for fragile patients. The results obtained in these patients were extremely encouraging.

The advantages and inconveniences of gastric endoscopy, radiography, and scintigraphy are discussed and compared.

Introduction

Peptic ulcer disease is common: 5% to 10% of all individuals develop an ulcer at some time during their lifetime [1].

The detection of gastric and duodenal ulcers was, up to a few years ago, the exclusive province of radiography and endoscopy. These two techniques are not mutually exclusive and, most often, are complementary.

Progressively, endoscopy has become the method used the most often, particularly in cases of hemorrhage [2].

Conventional isotopic techniques, which are relatively old, permit the diagnosis of gastroduodenal ulcers or gastritis when these lesions are bleeding

H.J. Biersack and P.H. Cox (eds), Nuclear Medicine in Gasteroenterology, 139–151

[3]. The principle of the technique involves the intravenous administration of red blood cells marked by Technetium 99m (^{99m}Tc) and their fixation on the lesion.

Recently, observations of non-bleeding gastritis have been reported, and the hyperfixation of the isotope is attributed to the hyperemia [3].

When there is no hemorrhaging, the method is not at all specific for gastroduodenal ulcers and is of no interest for their detection.

In 1983, a new method for detecting gastroduodenal ulcers was proposed [4, 5]. The basis of this method was the particular affinity of sucralfate for ulcerated gastric and duodenal tissues [6].

Sucralfate, an aluminium salt of a sulfated disaccharide, is a virtually unabsorbed drug and is thought to act locally at the site of an ulcer as a barrier to acid, pepsin, and bile, factors considered to be etiologic in the disease [6]. Sucralfate is considered safe and is as effective as cimetidine in the treatment of gastric and duodenal ulcers [7, 8]. Sucralfate has been shown to be cytoprotective in that it stimulates endogenous gastric prostaglandin production [9].

Vasquez et al. [4] have demonstrated the possibility of identifying, by means of marking the sucralfate with ^{99m}Tc, gastroduodenal ulcers induced in rabbits by acetylsalicylic acid in comparison with normal rabbits. They then studied 4 patients with gastroduodenal ulcers and 3 patients without ulcers. Scintigraphy was not positive except in the ulcer cases.

Braunstein et al. [5] confirmed these initial results and showed that the sole ulcer case that did not fix the marked sucralfate was that of a patient who was bleeding.

Subsequent to these publications, we conducted a preliminary study in 1987 with 23 patients with gastric or duodenal ulcers demonstrated by endoscopy and histology in all cases. In most of these cases, the location of the ulcer was confirmed by radiography [10].

We demonstrated that our technique supplied evidence of hyperfixation of the marker on the ulcer site in 18 cases out of 23 (sensitivity: 78%), including the 5 patients who were bleeding at the time of the examination.

The use of good technique is absolutely necessary, and it is essential to specify its limits and to compare it with other diagnostic methods.

We have the four following objectives:

1. to evaluate the *sensitivity* of our method in a larger number of patients and with a more refined technique;
2. to test the degree of *specificity* of the marking of the sucralfate for gastroduodenal ulcers;
3. to see if, in the post-treatment *follow-up* of patients with a formally demonstrated gastroduodenal ulcer, another endoscopy can be avoided and scintigraphy with sucralfate would suffice;
4. when an endoscopy and/or radiography are not feasible (for example, in

critically ill patients), and if there is a *high degree of suspicion of an ulcer* to see if scintigraphy is positive.

Material and methods

Marking the sucralfate

The marker is prepared as follows. We dissolve 3 gr of granulated sucralfate in an 8 ml solution of HCl (pH 4.3) and add to this mixture 0.05 to 0.07 mg of stannous tartrate. We then mark a kit of human serum albumin (HSA) with a solution of 50 mCi of ^{99m}Tc in 10 ml of distilled water.

We add 0.6 ml of the marked albumin solution, which is equivalent to 3 mg of HSA, to the sucralfate base. The activity of the tracer is thus 3 mCi.

We centrifuge this mixture at low speed for 10 min and then remove the supernatant fluid, which contains the technetiated HSA that is not fixated to the sucralfate.

We redissolve the centrifugation residue in 20 ml of distilled water. The marking yield is excellent, as the fixation percentage is from 98% to 99%. Thus, there remains in the supernatant fluid only 1% to 2% of the activity introduced into the mixture.

The product is stable for at least 5 h after its preparation.

Patients

The sensitivity of the method
This study concerns 64 cases of ulcers demonstrated by endoscopy and histology. In 20 cases, radiography confirmed the location of the ulcer. The age of the patients ranges from 20 to 90 years old (mean: 61 years old). Forty-one of them were men, and 23 women. Of the 64 patients, 6 were hemorrhaging during the examination. We studied separately the gastric ulcers, the duodenal ulcers, and those appearing after a partial gastrectomy.

The specificity of the method
This study concerns 31 patients in whom endoscopy and histology were normal or demonstrate another pathology than a gastric or duodenal ulcer. These patients were classified into three groups on the basis of the endoscopic and histological results:

- 10 normal patients (5 men, 5 women, average age: 63 years old);
- 15 cases of gastritis or bulbitis (9 men, 6 women, average age: 60 years old);
- 6 cases of cancer (2 men, 4 women, average age: 60 years old).

Follow-up
In 8 patients with gastroduodenal ulcers, endoscopy with histology and scintigraphy with marked sucralfate were performed before and after 8 weeks of treatment with cimetidine or ranitidine.

Scintigraphy with marked sucralfate without proof of a gastroduodenal ulcer
This was performed in 4 patients. Endoscopy or radiography were impossible (critically ill), but there was a history of ulcers in 4 cases and a strong suspicion of recidivity (typical pain in 2 cases, hematemesis in the other 2).

Imaging procedure

The patients had to fast for 12 h before the examination. If ulcer treatment was already in progress, we interrupted it at least 24 h before the examination.

The tracer was administered to the patient orally, and the container was rinsed with about 10 ml of distilled water. A half an hour after the administration of the tracer, we had the patient drink about 100 ml of distilled water to wash away residual ^{99m}Tc–HSA not fixed to the mucosa.

A half hour before the second series of images, we again had the patient drink 100 ml of distilled water. Whenever possible, the patient was asked to walk around a little in order to change the position of the stomach.

To make the images, we used a gamma camera with a data processing system (Siemens gammasonics camera and a computer).

We made the images one and two hours after the tracer was administered with the patient standing and lying down. The images were made with 200 000 and 400 000 bits depending on the fixation. When possible, we also placed reference points.

If the fixation remained unchanged on the second image, we concluded that it was an ulcerated image. However, it was difficult to wash away residual sucralfate in the gastric fundus, regardless of the position of the patient or the amount of liquid ingested. This makes interpretation in this area impractical.

Results

Sensitivity of the method in gastroduodenal ulcers

Gastric ulcers (35 patients)
The scintigraphy was positive in 29 out of the 35 patients and negative in 6. The sensitivity of the method is thus 82.8% for stomach ulcers. Scintigraphy

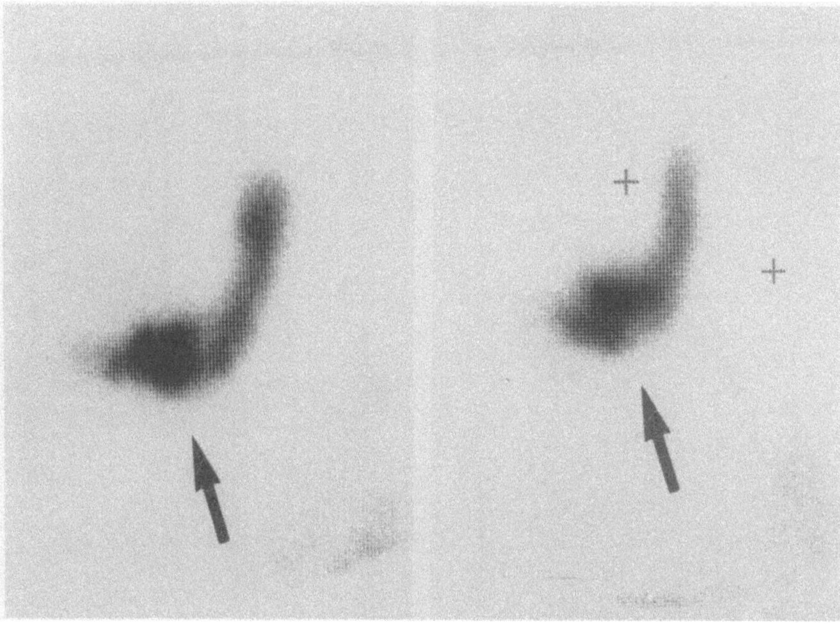

Figure 1. Ulcer of the stomach (arrows) demonstrated by scintigraphy on 1 h image (left) and 2 h image (right). Note retention in area of the ulcer on both 1 h and 2 h images. Uptake in fundus of stomach is also persistent although no pathology existed in this area.

was positive in the 3 patients who hemorrhaged during the examination. Figure 1 shows a gastric ulceration.

Duodenal ulcers (26 patients)

The scintigraphy was positive in 22 out of 26 patients and negative in the other 4. The sensitivity was thus 84.6% for duodenal ulcers. Scintigraphy was positive in the 3 patients who hemorrhaged during the examination. Figure 2 shows a duodenal ulceration.

Ulcers after partial gastrectomy (3 patients)

The scintigraphy was positive in all 3 cases. There was no hemorrhaging. Figure 3 shows an ulceration at the site of the gastro-jejunal suture. Figure 4 gives 3 ulcerations in the same patient. Table 1 summarizes the results concerning the sensitivity of the method for all the ulcer patients: 84.3% (54 positive out of 64 patients).

Specificity of the method

The 10 patients with a normal endoscopy and histology also had a normal scintigraphy with sucralfate. Of the 15 patients with gastritis or bulbitis, 12

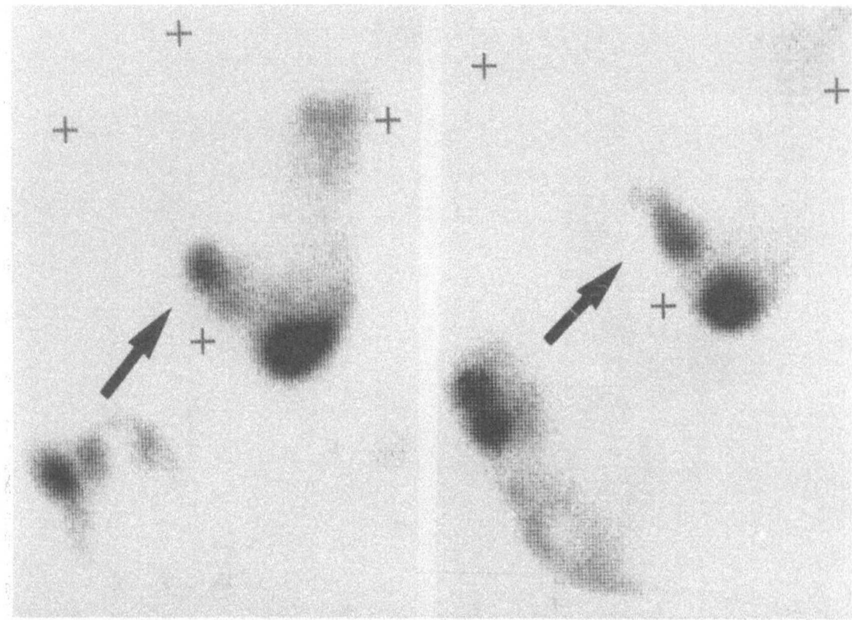

Figure 2. Bulbar ulcer of the duodenum (arrows). Left: 1 h image – Right: 2 h image. Again note retention of activity in both scintigrams and washout elsewhere except for the fundus of the stomach.

presented positive scintigraphy. The 6 cases of cancer were all positive with scintigraphy.

The specificity of the scintigraphy with sucralfate for gastroduodenal ulcers is determined as follows:

$$\text{Specificity} = U/T$$

where U is the number of positive examinations for gastroduodenal ulcers and T is the total number of positive scintigraphic examinations (ulcers, gastritis and bulbitis, cancer). Thus with $U = 54$ and $T = 54 + 12 + 6 = 72$, the specificity of the marking method for gastroduodenal ulcers is 54/72, i.e. 75%.

Follow-up

In the 6 cases in which the ulcer was cured, sucralfate scintigraphy was negative. In the 2 cases where the ulcer persisted, the scintigraphy was positive.

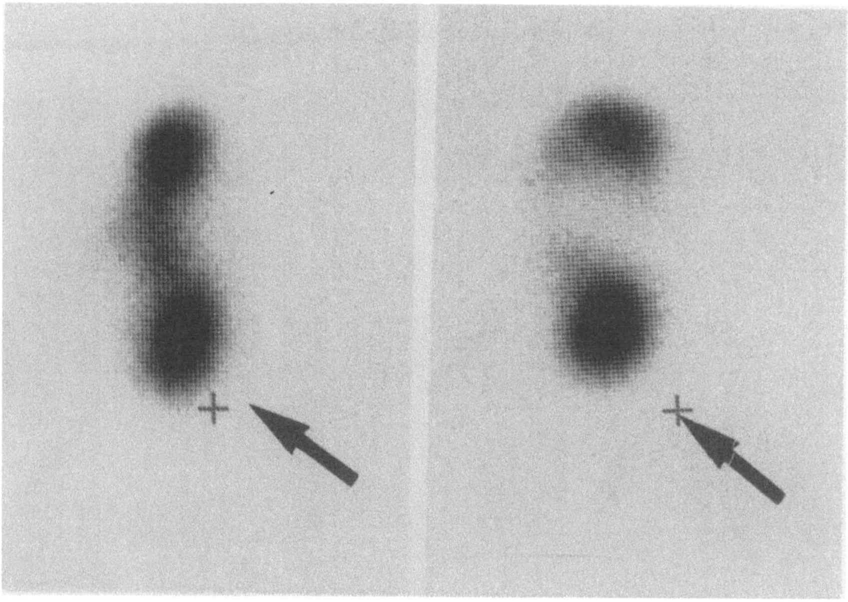

Figure 3. Ulcer of the gastrojejunal suture after partial gastrectomy (arrows) demonstrated by scintigraphy on 1 h image (left) and 2 h image (right). Again note retention of activity in the fundus of the stomach.

Scintigraphy alone in the cases of suspected ulcers

The examination was positive in all 4 cases, including the two cases with hematemesis.

Discussion

Our discussion will concern gastric scintigraphy with ^{99m}Tc-HSA-sucralfate and also the other techniques that are currently being proposed.

^{99m}Tc-HSA-sucralfate scintigram

We have shown in our preliminary study that the detection of gastroduodenal ulcers with ^{99m}Tc-HSA-sucralfate is easy and nontraumatic, and permits the screening of active mucosal ulcerations [10]. The examination was considered positive when the labeling agent accumulated significantly in the stomach, with the exception of the fundus, in the duodenum, and, in the cases of partial gastrectomy, at the site of the gastro-jejunal suture.

In 1987, the sensitivity of our method was 78% with 23 cases. At present,

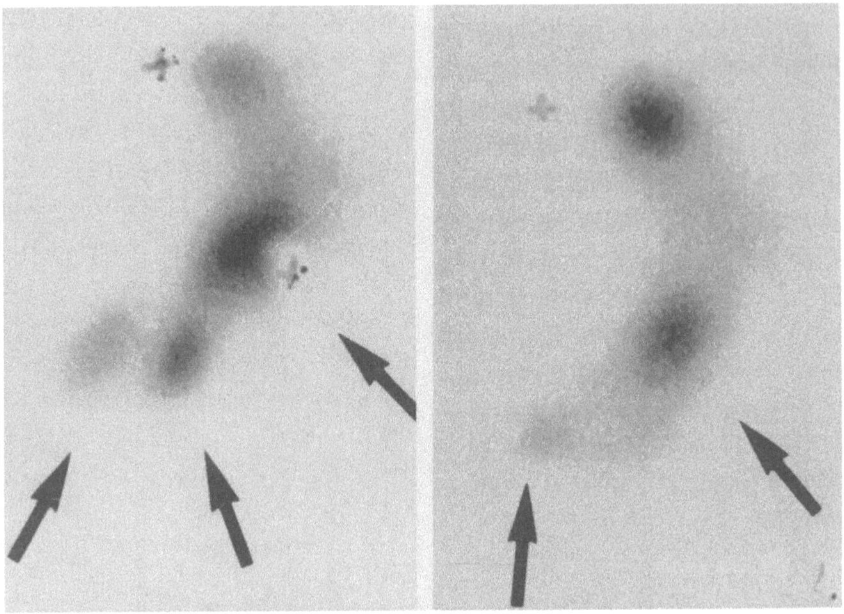

Figure 4. Two gastric and 1 bulbar ulcers (arrows) demonstrated by scintigraphy on the 1 h image (left). On the 2 h image (right), only 2 out of the 3 ulcers are positive. Again note retention of activity in the fundus of the stomach.

with a larger number of cases, the sensitivity is 84.3% for all of the ulcerous patients, including those who were hemorrhaging during the examination.

If the 3 cases of ulcers detected after a partial gastrectomy are excluded, the sensitivity of our method is not different for gastric ulcers (82.8%) and for duodenal ulcers (84.6%). Like other authors, we had no false-positive results [11, 12].

Our results as regards the sensitivity of a method of detection of gastroduodenal ulcers with sucralfate marked with ^{99m}Tc (84.6%) are of the same order as those of Dawson et al. (85.7%), but our series has 64 cases in contrast to their 14 [12]. Vasquez et al. obtained a sensitivity of 66% with 24 patients [11].

We and others thought it possible to avoid traumatic examinations like endoscopy in the identification of gastroduodenal ulcers [10, 12]. Since our

Table 1. Results concerning the sensitivity of the method for the ulcer patients

Ulcers	Number of cases	Positivity	Sensitivity
Gastric	35	29	82.8%
Duodenal	26	22	84.6%
After partial gastrectomy	3	3	100%
Total	64	54	84.3%

technique is only specific in 75% of the cases, one cannot be satisfied with scintigraphy to formally establish a gastroduodenal ulcer. Indeed, marked sucralfate fixates also in cases of gastritis, bulbitis, and cancer. We note that the idea of identifying cancers was suggested in 1983 by Vasquez et al. [4] and that this was demonstrated in 1 case by Dawson et al. in 1985 [12] and in 6 cases by ourselves.

As regards the follow-up of gastroduodenal ulcers, nevertheless, it would seem that a new endoscopy could be avoided in making the decision to continue or stop the anti-ulcer treatment.

For scintigraphy with sucralfate in patients for whom there is a strong suspicion of a gastroduodenal ulcer but for whom an endoscopy is impossible, the positivity of 4 cases out of 4 is interesting.

Of course, the proof of a benign ulcer is not provided, but it is probable in 75% of the cases, which makes the therapeutic decision of the clinician easier. In our experience, scintigraphy administered during a hemorrhage never modified the image, contrary to the initial observations of Braunstein et al. [5].

In view of the great frequency of gastroduodenal ulcers and the need for patients to submit either to radiography or endoscopy to establish the diagnosis, the question is whether gastroduodenal scintigraphy constitutes an attractive alternative method.

Thus it is useful to compare the advantages and inconveniences of the three exploratory methods.

Endoscopy, which permits taking histological samples, is the most sensitive method for the diagnosis of gastroduodenal ulcers. Dawson et al. estimated its sensitivity at 100% [12] and Dooley et al. at 92% [13]. By way of comparison, the sensitivity of radiography was found to be 75% by Dawson et al. [12] and 54% by Dooley et al. [13].

The sensitivity of scintigraphy with sucralfate is 84.3% in our experience with 64 patients. This sensitivity is estimated at 66% by Vasquez et al. with 24 patients [11] and at 85% by Dawson et al. for 14 patients [12]. Endoscopy, nevertheless, is the least comfortable of the methods. It frequently requires sedation, and there is a risk of perforation [11, 12]. It is also more onerous [14].

Radiographic techniques necessarily entail exposure to a significant amount of ionizing radiation, particularly since several images are often required [11]. For Pera et al., the radiographic procedure involves a radiation dose of around 0.8 mGy [14]. This dose seems to us to be very low, particularly in view of the dose of 2 to 3 mGy estimated by other authors [15].

There are several advantages of gastric scintigraphy relative to radiography. The technique is easily accepted by the patient, easy to repeat, and less expensive, and it provides evidence of ulceration of the duodenal bulb in zones deformed by cicatrices [12]. Finally, there is a lower calculated radiation dosage to the patient with ^{99m}Tc marked sucralfate than with a conventional barium meal [12]. This radiation calculated by us, is 1 mGy.

As regards specificity, there is no doubt that endoscopy and histology are the most effective.

Since an endoscopy is not always possible, it is interesting to compare the specificity of radiography and of scintigraphy for gastroduodenal ulcers. Dooley et al. estimate the specificity of radiography at 91% [13]. We have noted a specificity of 75% with scintigraphy with marked sucralfate.

It is possible that this percentage differs for a larger number of patients and with the diversification of the pathologies studied. The specificity estimated by Vasquez et al. [11] is 100%, but these authors did not take account of scintigraphic positivity with other pathologies. In addition, their study only concerned 24 patients.

The fixation of the marked sucralfate in other gastroduodenal pathologies than ulcers has been described by other authors than ourselves, and particularly for one case of stomach cancer [12]. That sucralfate fixes on ulcerated gastroduodenal tissues is well established [6]. Furthermore, in the intestines, sucralfate can fix on the mucous membrane after a polypectomy and in radiation proctitis when administered by enema [16, 17].

Other techniques than sucralfate labeling with ^{99m}Tc-HSA

^{14}C-sucralfate and ^{14}C-potassium sucrose sulfate

Sucralfate marked with ^{14}C has been compared with a solution of ^{14}C and potassium sucrose sulfate as a marker of ulcers induced in the rat [18]. It was considered that the sucrose sulfate moiety of sucralfate forms stable complexes with protein by means of a strong electrostatic interaction [19, 20]. In contrast with gastric acidity, a high concentration of sucrose sulfate is located at the proteinaceous surface of the ulcer [21].

It has turned out that ^{14}C-potassium sucrose sulfate leaves the gastric cavity one hour after administration with a fixation of only 2% on the mucous membrane, while the fixation of the sucralfate solution is more than 60% on the ulcerated mucous membrane in the same amount of time.

In addition, with an acid pH, the ^{14}C-labeled sucralfate forms an insoluble paste that adheres to the ulcerated mucous membrane, while the ^{14}C-potassium sucrose sulfate is soluble in water. These studies, which have not had any clinical application, have served as a point of departure for all the later studies and for the development of scintigraphy with gastroduodenal ulcers.

Selenium-75 (^{75}Se) and Indium-111m (^{111m}In)

In the experiment conducted by Knight et al., sucralfate was radiolabeled with ^{99m}Tc-HSA, ^{75}Se and ^{111m}In [22]. The three preparations were then administered to rats with gastric ulcers, and the gamma-camera images were compared. The authors then studied the radioactivity in vitro at the ulcer site after removing the stomachs of the rats. They concluded that, of the three gamma-emitting labels, ^{75}Se is the best marker for sucralfate and that

^{75}Se-sucralfate is suitable for studying the kinetics of sucralfate in human subjects.

Nevertheless, no clinical studies have yet been made on the matter, so the suggested superiority of ^{75}Se-sucralfate over the other markers has not been confirmed.

^{99m}Tc-HSA-potassium sucrose sulfate

On the basis of the work of Nagashima et al. [18–20], Vasquez et al. marked potassium sucrose sulfate with ^{99m}Tc-HSA [11]. They studied three patients with gastroduodenal ulcers with this marker and observed only one positive result. The small number of their patients prevented them from drawing any conclusions, but they considered that 'by this method, the background is probably reduced enough to enhance imaging of the ulcer'.

Marking the sucralfate in vivo

One study of the detection of gastric ulcers by marking the sucralfate in vivo was done by Pera et al., who compared it with the in vitro marking technique [14]. Indeed, they did not consider the preliminary results of Vasquez et al. very encouraging [4]. Their in-vivo marking technique consists of administering the sucralfate and a solution of stannous chloride orally and then, two hours later, a solution of sodium 99m-pertechnetate. This method was first tested on a model rabbit ulcer. They then studied 26 humans with sucralfate labeled directly in-vivo (15 normal subjects and 11 patients with a gastric ulcer). All the normal subjects gave true negative results, and all the ulcer patients true positive results (100%).

In comparison with their in-vitro study, in which they obtained 3 positive examinations out of 11 gastric-ulcer cases, i.e. 27%, the author concluded that the in-vivo marking method is preferable [14].

In addition, it involves fewer manipulations, and the patient receives much less radiation. We note that no cases of duodenal ulcers were studied.

Tietze et al. noted that, unlike the in-vitro technique, the in-vivo marking of sucralfate recommended by Pera et al. involves a risk [23]. In the event of slow gastric functioning, a portion of the free ^{99m}Tc does not fix on the sucralfate, but is captured by the thyroid and the quality of the gastric image can be poorer. For in-vivo marking, they suggest blocking the thyroid by sodium perchlorate, which they also administer with in-vitro marking.

Tietze et al., moreover, consider that the in-vitro technique gives just as good results as the in-vivo technique [23].

Seevers, a co-author of the article by Pera et al., did not find the thyroid radioactivity reported by Tietze et al. He offers no explanation for this difference [24].

Conclusions

The demonstration of gastroduodenal ulcers is easy to do at present with radioisotopic methods. The most common technique uses sucralfate marked with ^{99m}Tc-HSA. Other techniques, still being studied, seem to give poorer results, although they certainly are worth developing.

The sensitivity of scintigraphy with ^{99m}Tc-HSA-sucralfate for gastroduodenal ulcers is 84.3%, but the method has only a 75% specificity. Indeed, scintigraphic gastroduodenal images are also positive in cases of gastritis, bulbitis, and cancer.

Scintigraphy with sucralfate can, in some cases, be an interesting alternative method for endoscopy and radiography for the detection and the follow-up of gastroduodenal ulcers.

It is comparable in effectiveness to radiography but is less sensitive and less specific than endoscopy. Since endoscopy and radiography are not always possible, for example, with critically ill patients, and because endoscopy is not free of inconveniences, gastric scintigraphy seems to us to be a diagnostic method worthy of interest. In addition, it permits the visualization of an active ulcer in a cicatricial zone and is positive in the presence of hemorrhaging.

Acknowledgements

The authors would like to thank Dominique Denis for her secretarial assistance and Martine Mathieu, Aude Morel, Jacques Degardin, Geneviève Passerard, Hélène Roux for their technical assistance.

References

1. Schiller LR (1988) 'Peptic ulcer: epidemiology, clinical manifestations and diagnosis,' in: Wyngaarden JB, Smith Jr LH eds, *Cecil Textbook of Medicine*. 18th ed., p. 696, Philadelphia: W.B. Saunders Company.
2. Kahn KL, Kosecoff J, Chassin MR, Solomon DH, Brook RH (1988) 'The use and misuse of upper gastrointestinal endoscopy.' *Ann Intern Med* 109: 664–670.
3. Wilton GP, Wahl RL, Juni JE, Froelich JW (1984) 'Detection of gastritis by ^{99m}Tc-labeled-blood-cell scintigraphy.' *AJR* 143: 759–760.
4. Vasquez TE, Bridges RL, Braunstein P, Jansholt AL, Meshkinpour H (1983) 'Gastrointestinal ulcerations: detection using a Technetium-99m-labeled ulcer-avid agent (work in Progress).' *Radiology* 148: 227–231.
5. Braunstein P, Vasquez TE, Bridges RL, Jansholt AL, Meshkinpour H (1983) 'Tagged ulcer-avid material imaging (TUMI): a potent new method for the evaluation of peptic ulcer disease (abstract).' *J Nucl Med* 24: 78.
6. Martin F, Farley A, Gagnon M, Bensemana D (1982) 'Comparison of the healing capacities of Sucralfate and Cimetidine in the short-term treatment of duodenal ulcer: a double blind radomized trial.' *Gastroenterology* 82: 401–405.

7. Lahtinen J, Aukee S, Miettinen P, Poikolainen E, Pääkkönen M, Sandström R (1983) 'Sucralfate and Cimetidine for gastric ulcer.' *Scand J Gastroent* 18, suppl. 83: 49–51.
8. Hentschel E, Schütze K, Dufek W (1983) 'Controlled comparison of Sucralfate and Cimetidine in duodenal ulcer.' *Scand J Gastroent* 18, suppl. 83: 31–35.
9. Crampton JR, Gibbons LC, Rees N (1987) 'Effect of Sucralfate on Gastroduodenal bicarbonate secretion and prostaglandin E2 metabolism.' *Am J Med* 83: 14–18.
10. Puttemans N, Lambert M, Andre PP, Jamsin S, Balikdjian D, Lustman F (1987) 'Detection of Gastroduodenal ulcers using Technetium-99m-labeled sucralfate.' *J Nucl Med* **28**: 521–523.
11. Vasquez TE, Evans DG, Hartman MT, Hagan P, Fardi M, Ashburn WL (1986) 'Radionuclide imaging using Technetium-99m-labeled Sucralfate and Potassium sucrose sulfate to detect gastric and duodenal ulcers.' *J Nucl Med allied Sci* 30: 141–148.
12. Dawson DJ, Kahn AN, Nuttal P, Shreeve DR (1985) 'Technetium 99m-labelled sucralphate isotope scanning in the detection of peptic ulceration.' *Nucl Med Commun* 6: 319–325.
13. Dooley CP, Larson AW, Stace NH, Renner IG, Valenzuela JE, Eliasoph J, Coletti PM, Halls JM, Weiner JM (1984) 'Double-contrast barium meal and upper gastro-intestinal endoscopy. A comparative study.' *Ann Intern Med* 101: 538–545.
14. Pera A, Seevers RH, Meyer K, Hall C, Bekerman C, Anderson TM, Katzen H, Laakso L, Pinsky SM (1985) 'Gastric ulcer localization by direct in vivo labeling of sucralfate.' *Radiology* 156: 783–786.
15. Commission Radiodiagnostic. Société Française des Physiciens d'Hôpital. Evaluation des doses délivrées au cours d'examens radiologiques. Rapport préliminaire. Mai 1988.
16. Bronner MH, Yantis PL (1986) 'Intracolonic Sucralfate suspension for postpolypectomy hemorrage.' *Gastrointest Endosc* 32: 362–363.
17. Kochhar R, Sharma SC, Gupta BB, Mehta SK (1988) 'Rectal Sucralfate in radiation proctitis.' *Lancet* ii: 400.
18. Nagashima R, Hinohara Y, Tohira Y, Kamiyama H (1980) 'Selective binding of Sucralfate to ulcer lesion. II. Experiments in rats with gastric ulcer receiving ^{14}C Sucralfate or potassium ^{14}C Sucrose sulfate.' *Arzneim Forsch* 30: 84–88.
19. Nagashima R, Yoshida H (1979) 'Sucralfate, a basic aluminium salt of sucrose sulfate. I. Behaviors in gastroduodenal pH.' *Arzneim Forsch* 29: 1668–1676.
20. Nagashima R (1981) 'Mechanisms of action of sucralfate.' *J Clin Gastroenterol* 3 (suppl 2): 117–127.
21. Nakazawa S, Nagashima R, Samloff IM (1981) 'Selective binding of sucralfate to gastric ulcers in man.' *Dig Dis Sci* 26: 297–300.
22. Knight LC, Fischer RS, Malmud LS (1984) 'Comparison of labels for carafate in a gastric ulcer model (abstract).' *J Nucl Med* 25: 119.
23. Tietze P, Gaatz KF, Hundeshagen H (1986) 'Gastric ulcer localization by direct in vivo labeling of Sucralfate.' *Radiology* 161: 569.
24. Seevers RH (1986) 'Gastric ulcer localization by direct in vivo labeling of Sucralfate.' *Radiology* 161: 569–570.

10. Gastroesophageal and Biliary Reflux

ROLAND BARES and UDALRICH BUELL

Introduction

Within the past 15 years an increasing number of technical procedures have been developed to diagnose gastroesophageal or duodenogastric reflux (Table 1). They are aimed at measuring different features of reflux such as retroperistaltic transport of food or digesta, insufficient function of anatomic sphincters, or changing composition of gastric or esophageal contents. Among them nuclear medicine procedures have gained importance because of their easy handling and non-invasive way of investigation. After a brief recapitulation of reflux pathogenesis, this review will present the state of the art of nuclear medicine techniques as well as a summary of clinical results and considerations about indication.

Pathophysiology

The main reason for gastroesophageal reflux is incompetence of the lower esophageal sphincter caused by either inadequate function of the sphincter or by intrathoracic position of the sphincter system (presbyesophagus, hiatus hernia) [7, 8]. Depending on its composition (pH, presence of bile acids) and duration gastroesophageal reflux may lead to clinical symptoms such as regurgitation, retrosternal burn, or epigastric pain, which are often accompanied by moderate to severe esophagitis [12, 33, 43, 44]. On the other hand, there is no strict correlation between clinical symptoms and histological evidence of esophagitis [27]. Thus, a precise diagnosis is most important to select adequate therapy (drugs vs. surgery).

Duodenogastric reflux occurs due to malfunction of the pylorus and gastric antrum [40]. Like with gastroesophageal reflux its effect depends on composition (pH, quantity of bile acids and pancreatic enzymes) and reflux duration. Small amounts of biliary reflux which are rapidly cleared from the stomach are a physiologic finding. Larger amounts of bile combined with acid gastric

H.J. Biersack and P.H. Cox (eds), Nuclear Medicine in Gasteroenterology, 153–167

Table 1. Diagnosis of gastroesophageal and duodenogastric reflux: technical procedures

Gastroesophageal Reflux
X-ray (fluoroscopy following barium meal)
Esophago-gastroscopy
Gastroesophageal reflux scintigraphy
24 h pH-probe monitoring
Manometry
Duodenogastric Reflux
X-ray (fluoroscopy following barium meal)
Gastro-duodenoscopy
Hepatobiliary reflux scintigraphy
pH-probe monitoring
Bile acid measurements in gastric aspirates

contents, however, may lead to severe gastritis or ulceration [1, 42]. Even alkaline reflux esophagitis can arise, if additional gastroesophageal reflux is present [33, 43]. There are no symptoms, which are specific for duodenogastric reflux. In most of the patients they are attributable to coexistent gastric pathologies (ulcers, motility disorders) [45]. So evidence of duodenogastric reflux has to be interpreted carefully in respect to therapeutic consequences.

Methods

Gastroesophageal reflux scintigraphy

Detection and measurement of gastroesophageal reflux by nuclear medicine procedures is based on radiolabelling of gastric contents. The technique was firstly described by Fisher et coworkers in 1976 [14] and has become far spread over US and Europe [24, 29, 31, 32].

The technical equipment which is necessary to perform gastroesophageal scintigraphy is simple and should be available in each nuclear medicine department:

1. A digital gamma camera equipped with high sensitivity (LFOV camera) or diverging collimator and interfaced to a dedicated computer system for digital data storage and processing.
2. An inflatable abdominal binder for controlled reflux provocation [14].

Suitable radiopharmaceuticals have to be stable in acid media and neither mucosa adhesive nor permeable. Both ^{99m}Tc labelled DTPA and sulfur colloid fulfil these requirements. A dose of 20–40 MBq is sufficient to gain reliable scintigraphic results. Radiation exposure from gastrointestinal scintigraphy is low (total body 0.005 cGy/10 MBq) and several times less than that from barium esophagogastrography [46, 49].

Scintigraphy is usually performed after a 12 h (overnight) fasting period.

The patient swallows the radioactive compound mixed with 200 ml water or, in an effort to increase reflux incidence, with diluted lemon juice or hydrochloric acid (0.1 M) to acify the gastric contents. After the inflatable binder is wrapped around his abdomen he is positioned supine under the gamma camera so that the gastric fundus is delineated in the lower part of the camera field of view. Reflux provocation is achieved by gradually inflating the binder to increase the lower esophageal sphincter gradient [14, 30]. In addition physiologic provocation by Valsalva manouver or coughing may be employed.

Analog images are obtained at 30 s intervals for 240 s beginning at rest and proceeding parallel with increasing pressure of the abdominal binder. Digital acquisition covers also 240 s with a frame rate of 1 frame/2 s. Presence of reflux is evaluated visually and quantitatively by ROI-technique, drawing irregular ROIs over esophagus, gastric fundus and paraesophageal background. For each frame the amount of reflux is computed as background corrected ratio of esophageal to gastric radioactivity (Reflux Index [14], Equation 1). In reflux-free patients reported R.I. values range from 0.1 to 4.9% [14, 24, 29–32] depending on varying background correction. Figure 1 shows an example of gastrointestinal reflux in a patient with hiatus hernia and esophagitis grade 2. Table 2 summarizes the method of gastroesophageal scintigraphy.

$$RI\,(\%) = \frac{(\text{esophagus}) - (BGR)}{(\text{gastric fundus})} \times 100 \qquad (1)$$

A slightly modified gastroesophageal scintigraphy can be applied to infants and children: 2–4 MBq of the radiolabelled compound (whole body radiation exposure 0.006–0.02 cGy/4 MBq [19]) are given orally, preferably being mixed with formula milk (50–200 ml). The young patients are positioned supine under or above the gamma camera. Scintigrams are obtained for 60–90 min at 1 min intervals. Standard reflux provocation maneuvers do not exist. In elder children the above described abdominal binder may be used, in infants and younger children manual reflux provocation can be carried out if necessary. At the end of each study anterior and posterior scintigrams of the thoracic area should be obtained to detect possible pulmonary aspiration. Reflux quantification is similar to the method in adults although the normal range appears to be lower (effect of missing provocation?).

Complementary information about esophageal motor function and clearance becomes available, if esophageal transit is recorded [25]. By dividing the radioactivity into 2 portions, studies in upright and supine position can be performed. The gamma camera has to be positioned in front of the patient covering the complete thoracic area. Data acquisition is started immediately before swallowing the radioactivity mixed to a small volume of water or milk. Analog images are obtained in 4 s intervals and digital in

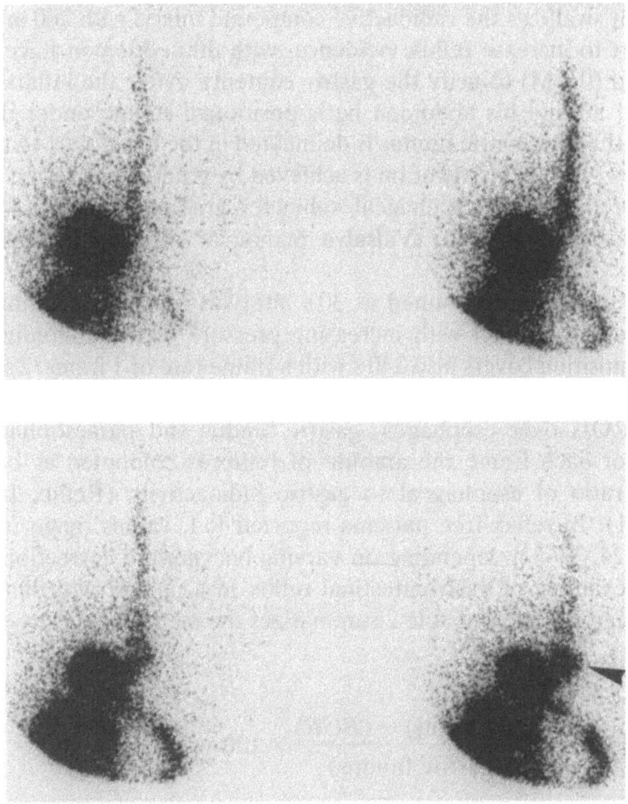

Figure 1. Gastroesophageal reflux scintigraphy in a 37 y old female patient with roentgenologically proven hiatus hernia (arrowhead) and grade 2 esophagitis demonstrating continuous reflux (reflux index = 8.7%).

1 sec intervals for 60 s. Esophageal transit can be quantified by computing esophageal clearance (Equation 2) using an irregular region of interest of the esophagus. In healthy patients values of esophageal clearance range between 85 and 95% [4, 5, 29]. Figure 2 demonstrates scintigraphic findings

Table 2. Method of gastroesophageal scintigraphy

Gamma camera	Digital, high sensitivity (LFOV) or diverging collimator
Radiopharmaceutical	^{99m}Tc labelled DTPA or sulfur colloid
Radioactivity	20–40 MBq
Application	Orally, mixed with 100–200 ml water
Acquisition	Anterior view, patient supine, analog images at 30 s intervals, digital storage 1 frame/2 s, in total 240 s, 64 × 64 matrix
Provocation	Inflatable abdominal binder, Valsalva maneuver
Processing	ROIs of esophagus, paraesophageal background, and gastric fundus, computing of RI (normal < 5%)

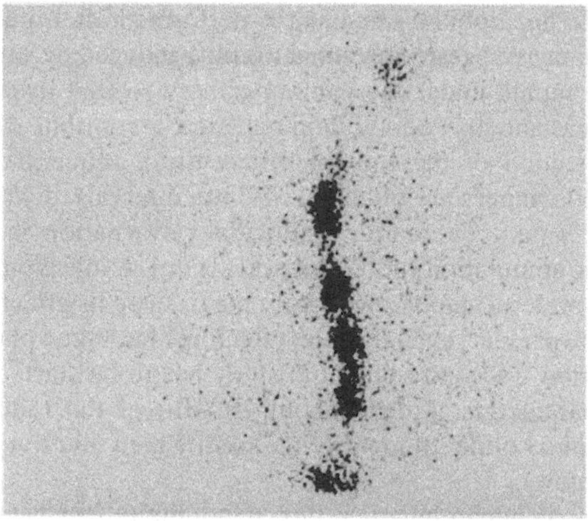

Figure 2. Disturbed esophageal motor function in a patient suffering from grade II esophagitis. Note the distinct scintigraphic pattern indicating disordered peristalsis.

in a patient suffering from grade II esophagitis. Lately special processing techniques have been developed to create parametric images to evaluate esophageal function [47]. More detailed information about this technique can be found in [15, 26].

$$\mathrm{ECl} = \frac{\mathrm{Cts}(T_{\max}) - \mathrm{Cts}(T_{\max} + 10)}{\mathrm{Cts}(T_{\max})} \times 100 \qquad (2)$$

Hepatobiliary reflux scintigraphy

Detection of duodenogastric reflux by nuclear medicine procedures is based on radiolabelling of bile. This can be achieved by i.v. injection of hepatotropic radiopharmaceuticals (IDA derivatives), which are rapidly excreted into bile following hepatocellulary transit [6]. Depending on the pharmakokinetics of the chosen compound, bile flow can be studied for a varying period of time [52].

Like with gastroesophageal scintigraphy the technical equipment, which is necessary to perform hepatobiliary reflux scintigraphy, is simple: a digital LFOV gamma camera interfaced to a dedicated computer system for data storage and processing. A large variety of suitable compounds to be labeled with ^{99m}Tc has been developed [6]. Differing features are velocity of blood clearance and hepatocellulary transit resulting in different labelling periods of bile. Because of limited camera imaging time compounds with fast hepatic transit are mainly used in clinical trials. A dose of at least 50 MBq (better 150–200 MBq) should be injected to gain sufficient countrates per frame. Whole body radiation exposure is about 0.082 cGy/185 MBq [46].

In principle hepatobiliary imaging is performed after a 6 h fasting period to exclude changes of gastrointestinal motility induced by eating. The patient is positioned supine under the gamma camera so that liver, the lower part of the thoracic and the mid-abdominal area are within the field of view. After i.v. injection of the radiopharmaceutical, sequential scintigraphy is started. Analog images are obtained at 2 min intervals, digital storage at 15 s intervals for 30 min. Thereafter, gallbladder evacuation should be induced by physiologic stimulation (milk, fatty meal) or i.v. injection of cholecystokinin [28]. We prefer a standardized fatty meal, since its effects are physiological and may be present, even if cholecystectomy has taken place. Immediately after stimulation a second sequential study has to be started using the same acquisition parameters. Finally about 10 MBq of the radiopharmaceutical should be applied orally to identify the gastric area and to quantify eventual reflux (see below).

Evaluation of hepatobiliary reflux scintigraphy can be done visually in respect of radioactivity within the gastric (and esophageal) area and quantitatively by ROI-technique. Various methods have been described differing from each other in background correction and in the way of quantification. Tolin and coworkers [51] divided the increase of gastric radioactivity by the decrease of radioactivity in the hepatobiliary tree for a given period of time (Equation 3). Most recently, a very sophisticated modification of this procedure using 6 different ROIs was developed by Dufresne [10]. Hyödynmaa et al. [22] published an easy technique based on calibrating gastric radioactivity by oral application of a known amount of radioactivity.

$$\mathrm{DGR} = \frac{\mathrm{S}_t - \mathrm{S}_0}{\mathrm{HB}_0 - \mathrm{HB}_t} \times 100 \tag{3}$$

(DGR = duodenogastric reflux (%); S = stomach; HB = hepatobiliary tree.)

The methods mentioned above have advantages and shortcomings. Correlating reflux with the total radioactivity leaving the hepatobiliary tree is highly dependent on absence of any pathology within liver, gallbladder, or bile ducts. Correlation with total intestinal radioactivity needs correct background subtraction in areas surrounding the liver (stomach!). Correlating gastric reflux with the injected radioactivity [22] is easy to perform, but like method 2 it depends highly on correct background subtraction and correction for superimposing bowel loops. In addition, quantification should also take into account reflux duration, which is only achieved by the method of Dufresne [10]. Figure 3 demonstrates biliary duodenogastric reflux in a patient with gastric ulcer, Table 3 summarizes the method of hepatobiliary reflux scintigraphy.

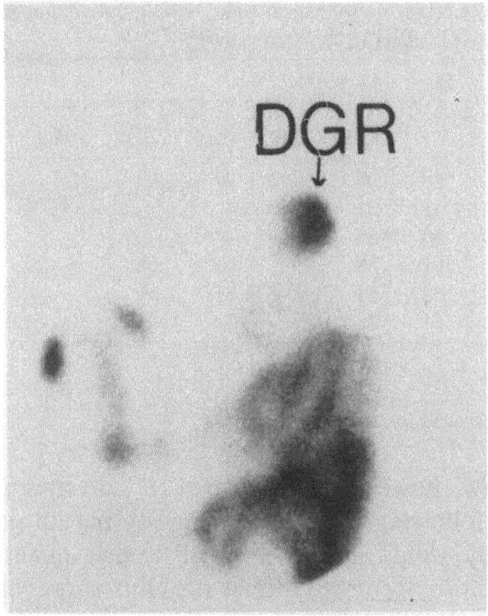

Figure 3. Hepatobiliary reflux scintigraphy presenting biliary duodenogastric reflux (arrow) in a 45 y old patient with gastric ulcer.

Table 3. Method of hepatobiliary reflux scintigraphy

Gamma camera	Digital, LEAP or high resolution collimator
Radiopharmaceutical	^{99m}Tc labelled IDA derivates
Radioactivity	50–200 MBq
Application	i.v.
Acquisition	Anterior view, patient supine, analog images at 2-min intervals, digital storage 1 frame/15 s for 2 × 30 min, 64 × 64 matrix, 1 single static image following gastric labelling (10 MBq), parameters see above
Processing	ROI-technique depending on method chosen

Results

Gastroesophageal reflux

Gastroesophageal reflux scintigraphy has been used in various clinical studies comparing its effectiveness with other technical procedures like 24 h pH-probe monitoring, manometry, endoscopy including histology, and X-ray techniques. A summary is given in Table 4. The studies are somewhat differing in respect to the accepted diagnostic 'gold standard,' but it can be stated that scintigraphy is more sensitive than endoscopy or X-ray investigations and little less than pH-probe monitoring. Validated results from patients who previously underwent gastric surgery are not available. Our

Table 4. Clinical results of gastroesophageal reflux scintigraphy compared with other technical procedures

Author	Year	Pts.	Sensitivity (%)						'Gold standard'
			GES	pH	End[1]	Hist[1]	Mano	*X*-ray[2]	
Fisher [14]	1976	30	90	–	40	47	77	60	symptoms
Leisner [29]	1978	51	88	100	–	–	65	58	24 h-pH
Martins [31]	1984	30	85	100	–	–	–	–	24 h-pH
Kaul [24]	1986	101	86	–	68	58	–	–	symptoms
Bares [5]	1987	32	69	–	31	31	–	47	symptoms
Bares[3]	1989	51	85	–	12	33	–	62	symptoms

[1]Esophagitis.
[2]Hiatus hernia.
[3]Patients following gastrectomy; unpublished data.

own data [4] show, however, that gastro- or enteroesophageal reflux is a frequent finding following complete (36%) or partial gastrectomy (20%), even if endoscopy yields normal results. Reflux quantification has been proven to be useful to document effects of reflux surgery or medical therapy [30].

Comparative reflux studies in infants and children [2, 19, 38] yielded similar results (sensitivity of gastroesophageal reflux scintigraphy 57–76% compared with 40/50% of manometry/X-ray resp.; 'gold standard' = 24 h pH-probe monitoring). In addition they could occasionally demonstrate concomitant aspiration detected by scintigraphy [38].

Disturbances of esophageal motor function are a frequent finding in patients presenting esophagitis or reflux disease. Tolin et al. [50] found diminished esophageal transit in patients with symptomatic gastroesophageal reflux and stated excellent correlation with manometric findings. Leisner [29] reported decreased esophageal clearance in about 50% of patients suffering from esophagitis. In general clinically apparent esophagitis (stage II–III) was accompanied by hampered esophageal motility.

Duodenogastric reflux

The importance of biliary duodenogastric reflux is one of the most controversely discussed topics in gastroenterology. Numerous clinical or experimental studies [1, 20, 21, 41, 42] have been performed to investigate the effects of bile salts on gastric mucosa using endoscopical or blind aspiration techniques. Scintigraphic results are summarized in Table 5. Like with the other techniques mentioned above the findings are not consistent and difficult to interprete, since different data processing and acquisition techniques have been employed. The clinical value of scintigraphic reflux quantification is still under discussion.

The relevance of biliary duodenogastric reflux following gastric surgery (vagotomy, partial gastric resection) is also subject to discussion. Hepatobili-

Table 5. Findings of hepatobiliary reflux scintigraphy in patients suffering from gastric or duodenal ulcerations and symptom-free controls

Author	Year	Pts.	Diagnosis	Duodenogastric reflux	(%)
Mosiman	1983	50	GU	34/50	68
		42	control	31/42	74
Dumont	1984	64	GU	34/64	53
		55	control	20/55	36
Wolverson	1984	30	GU	16/30	53
		15	control	8/15	53
Niemela	1984	33	GU	27/33	82
		33	control	14/33	42
Thomas	1984	14	GU	12/14	86
		15	DU	11/15	73
		20	control	4/20	20
Bares	1987	21	control	10/21	43
Dufresne	1988	20	DGR	13/20	65
		16	DU	7/16	61
		7	control	0.7	0

GU = gastric ulceration.
DU = duodenal ulceration.
DGR = duodeno-gastric reflux gastropathy [10].

Table 6. Frequency of duodenogastric reflux following gastric surgery demonstrated by hepatobiliary reflux scintigraphy

Author	Year	Pts.	Diagnosis	Duodenogastric reflux	(%)
Reichelt [39]	1977	42	B-I	19/42	46
		28	SPV + PP	11/28	40
		26	SPV − PP	3/26	11
Muhammed [36]	1980	17	TV + PP	8/17	47
		13	SPV	4/13	31
		15	B-I/B-II	15/15	100
Heidenreich [18]	1982	42	B-II	21/42	50
		50	SPV + PP	19/50	38
Karlqvist [23]	1983	11	B-I	11/11	100
		16	Roux-en-Y	4/16	25
Dumont [11]	1984	14	B-I	12/14	86
		26	B-II	26/26	100
Bares [5]	1987	7	B-I	6/7	86
		5	B-II	3/5	60

B-I = Bilroth-I hemigastrectomy (termino-terminal gastroduodenstomy).
B-II = Bilroth-II hemigastrectomy (termino-lateral gastrojejunostomy).
Roux-en-Y = Hemigastrectomy and bile diversion by termino-lateral (Y) jejunojejunostomy.
SPV = selective proximal vagotomy.
PP = pyloroplasty.

ary reflux scintigraphy has been used frequently to investigate changes of bile flow after different surgical procedures and in patients suffering from postoperative syndromes (afferent loop, dumping syndrome, alkaline reflux gastritis). Table 6 shows results following vagotomy and hemigastrectomy.

Table 7. Results of hepatobiliary reflux scintigraphy in completely gastrectomized patients

Author	Year	Pts.	Diagnosis	Biliary reflux (%) into	
				Substitute	Esophagus
Donovan [7]	1982	6	Omega	83	83
		20	Roux-en-Y	25	25
Gratz [16]	1983	10	Omega	100	30
		5	Roux-en-Y	80	20
		62	Interpos.	31	0
Gratz [17]	1984	4	Omega	100	100
		5	Roux-en-Y	0	0
		29	Interpos.	24	7
Bares [3, 5]	1987	23	Interpos.	52	13
		17	Roux-en-Y	17	6

Omega = termino-lateral esophago-jejunostomy.
Interpos. = gastric substitution by jejunal interposition.
Roux-en-Y = esophago-jejunostomy and bile diversion by jejuno-jejunostomy (Y).

Following complete gastrectomy hepatobiliary reflux scintigraphy is an accepted tool to detect biliary enteroesophageal reflux. It may cause alkaline reflux esophagitis which is a frequent postsurgical complication. Table 7 gives a summary of the results of reflux studies in gastrectomized patients.

Clinical validity

Gastroesophageal reflux scintigraphy has been shown to be highly sensitive in reflux-detection, even if no esophagitis is present. Like with 24 h pH-probe monitoring it may be argued whether reflux is of any clinical relevance, if it is not combined with esophagitis. A lot of clinical work has been done on this subject, and it has been proven that 'reflux disease' exists as clinical entity without esophagitis being present [13, 27]. It has also elucidated that reflux may gain importance, if the esophageal clearing function is hampered (i.e. if refluxed material contaminates the esophageal mucosa for longer periods of time). Thus, 'reflux disease' should be diagnosed, if a significant amount of reflux is present *and* esophageal motor function is disturbed. Scintigraphic measurement of esophageal transit and reflux supplies all necessary information to differentiate pathological and physiological refluxes.

On the other hand, 24 h pH-probe monitoring is more sensitive in reflux detection, able to measure reflux duration, and even to survey long periods of time, which cannot be done by scintigraphy. Thus, is gastroesophageal reflux scintigraphy still necessary? If endoscopy and 24 h pH-probe monitoring are available, then the answer has to be no! If not, or if the results are

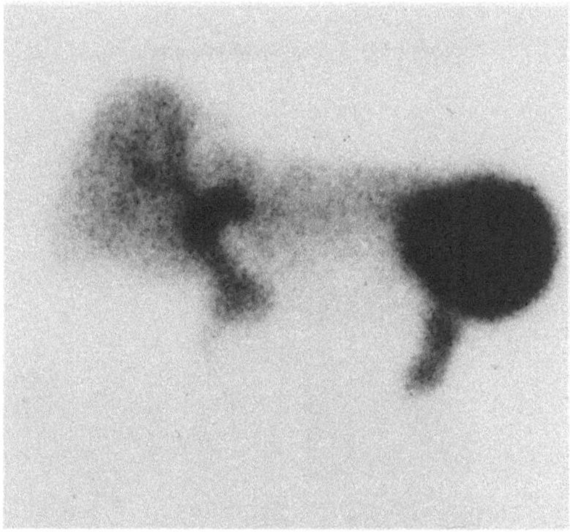

Figure 4. Biliary enterogastric reflux following Bilroth-II hemigastrectomy.

doubtful though clinical symptoms are highly suggestive for 'reflux disease,' then gastroesophageal reflux scintigraphy should be performed to establish the correct diagnosis.

Following previous gastric surgery 24 h pH-probe monitoring is of limited value, since gastric acid output may be diminished or completely absent (gastric replacement). So gastroesophageal scintigraphy might be the most sensitive diagnostic procedure. However, pathophysiological significance of non-acid reflux is low. In patients following complete gastrectomy we found frequently entero-esophageal reflux though neither clinical symptoms nor endoscopical or histological evidence of esophagitis were present. The only exception is biliary esophageal reflux which is proven to have a high mucosa-damaging potential [33, 43, 44].

Detection and clinical importance of duodenogastric reflux is even more under discussion than that of gastroesophageal reflux. It is a physiological finding in most of the healthy individuals [35], but may be relevant, if gastric mucosa is contaminated for longer periods of time such as in states of gastroparesis. Most of the clinical trials performing hepatobiliary reflux scintigraphy revealed no significant differences in healthy individuals and patients with gastric or duodenal ulcerations [11, 34, 53]. Studies using gastric aspiration techniques yielded similar results although individual findings did not correlate [20, 21, 35]. Thus, scintigraphic detection of biliary duodenogastric reflux seems to be of little diagnostic value. The importance of reflux quantification has not been proven yet, although some authors found significant differences between controls and ulcer patients [10, 37, 48].

In patients who had previously undergone gastric surgery duodenogastric

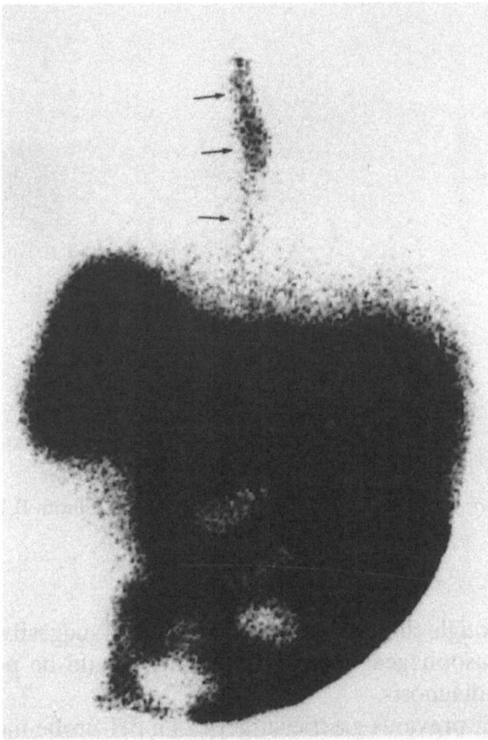

Figure 5. Biliary enteroesophageal reflux (arrows) suggestive for esophagitis (endoscopically proven) in a patient following complete gastrectomy demonstrated by hepatobiliary reflux scintigraphy.

reflux may also be physiological or even a necessary consequence of surgery (such as in the case of hemigastrectomy with Billroth-II anastomosis, Fig. 4). Alkaline reflux gastritis, a somewhat unclear clinical entity, is said to be caused by refluxing bile and/or pancreatic enzymes following hemigastrectomy [1, 41]. Up to now the exact pathophysiology could not be explained and bile diversion surgery did often fail to improve clinical symptoms. If complete gastrectomy has been performed, biliary reflux into the gastric substitute cannot be regarded pathological, but forms a precondition for biliary esophageal reflux which has a definite mucosa damaging potential as stated above. So hepatobiliary reflux scintigraphy is most useful in gastrectomized patients presenting symptoms of 'reflux disease' to select adequate therapy (drugs vs. surgery). In Fig. 5 scintigraphy demonstrates biliary enteroesophageal reflux in a patient presenting esophagitis following complete gastrectomy. The indications of gastroesophageal and hepatobiliary reflux scintigraphy are summarized in Table 8.

Table 8. Indications for gastroesophageal (GES) and hepatobiliary reflux (HBS) scintigraphy

1) Suggested gastroesophageal reflux	
24 h pH-monitoring not available or unconclusive	-GES-
Patients following previous gastric surgery	-GES + HBS-
Patients following complete gastrectomy	-HBS-
2) Gastric or duodenal ulcers	
Recurrent disease (prae and post surgery)	-HBS- (?)

Conclusion

Gastroesophageal reflux scintigraphy is a non-invasive diagnostic procedure which is based on radiolabelling of the gastric contents. Like with hepatobiliary scintigraphy it is easy to perform and should be available in each nuclear medicine department. Its sensitivity equals that of 24 h pH-probe monitoring so that it might be performed if pH-monitoring is not available or yielding inconclusive results. Hepatobiliary reflux scintigraphy is a unique diagnostic technique to demonstrate bile flow within the intestine. It is most important in post-surgical states to detect esophageal reflux causing alkaline reflux esophagitis. The relevance of biliary duodenogastric reflux is still open and should be subject to further clinical studies.

References

1. Alexander-Williams J (1982) 'Alkaline reflux gastritis: a myth or a disease?' *Am J Surg* 143: 17–21.
2. Arasu TS, Wyllie R (1980) 'Gastroesophageal reflux in infants and children – comparative accuracy of diagnostic methods.' *J Pediatr* 5: 798–803.
3. Bares R, Fass J, Buell U, Kleinhans E, Schumpelick V (1986) 'Changes of gastrointestinal motility after complete gastrectomy.' *Nucl Med* 4: A121.
4. Bares R, Fass J, Weiller G, Kleinhans E, Büll U, Schumpelick V (1988) 'Nuklearmedizinische Diagnostik funktioneller Veränderungen nach resezierenden Magenoperationen.' *Nucl Med* 2: 17.
5. Bares R (1988) 'Nuklearmedizinische Refluxdiagnostik.' *Nuklearmediziner* 5, 11: 331–7.
6. Chervu IR, Nunn AD, Loberg MD (1982) 'Radiopharmaceuticals for hepatobiliary imaging.' *Sem Nucl Med* 12: 5–17.
7 Clark J, De Meester TR, Johnson L, Skinner DB (1975) '24 hr lower esophageal pH-monitoring and the lower esophageal sphincter.' *Surg Forum* 26: 362–3.
8. Dodds WJ, Hogan WJ, Miller WN (1976) 'Reflux esophagitis.' *Am J Dig Dis* 21: 49–67.
9. Donovan IA, Fielding JWL, Bradby H, Sorgi M, Harding LK (1982) 'Bile diversion after total gastrectomy.' *Br J Surg* 69: 389–90.
10. Dufresne F, Carrier L, Gagnon M, Picard D, Chartrand R, Dumont A (1988) 'Scintigraphic study of duodenal-gastric reflux in cases of primary gastropathy, chronic ulcer of the duodenal bulb, and Moynihan's disease.' *J. Nucl Med* 29: 17–22.
11. Dumont A, Abramovici J, Barbier P (1984) 'Quantitative analysis of Diethyl-IDA reflux – comparison of controls, dyspeptic, gastric ulcerated and gastrectomized patients.' *Scand J Gastroenterol* 19 (suppl. 92): 206–9.
12. Evander A, Little G, Ridell RH, Walther B, Skinner DB (1987) 'Composition of the

refluxed material determines the degree of reflux esophagitis in the dog.' *Gastroenterology* 93: 280–6.
13. Fahrländer H (1981) 'Pathophysiologie, Klinik und Therapie der gastroösophagealen Refluxkrankheit.' *Schw Med Wschr* 111: 550–5.
14. Fisher RS, Malmud LS, Roberts GS, Lobis IF (1976) 'Gastroesophageal scintiscanning to detect and quantitate GE reflux.' *Gastroenterology* 3: 301–8.
15. Gibson CJ, Bateson MC (1985) 'A parametric image technique for the assessment of esophageal function.' *Nucl Med Commun* 6: 83–9.
16. Gratz KF, Creutzig H, Meyer HJ, Schober O, Pichlmayr R, Hundeshagen H (1982) 'Nachweis von Gallerreflux nach Gastrektomie mit der hepatobiliären Sequenzszintigrafie.' *RöFo* 137,4: 434–39.
17. Gratz KF, Meyer HJ, Creutzig H, Pichlmayr R, Hundeshagen H (1984) 'Nuklearmedizinischer Nachweis des ösophagealen Gallerefluxes.' *Nucl Med* 24: 35–8.
18. Heidenreich P, Vogt P, Eisenberger AS, Wölfle KD (1982) 'Diagnostik des duodenogastralen Reflux mit 99m-Tc-HIDA. Methode und Ergebnisse nach selektiv-proximaler Vagotomie (SPV) und Pyloroplastik.' *Nuklearmedizinier* 2,5: 83–94.
19. Heyman S, Kirkpatrick JA, Winter HS, Treves S (1979) 'An improved radionuclide method for the diagnosis of gastroesophageal reflux and aspiration in children (milk scan).' *Radiology* 131: 479–82.
20. Hoare AM, McLeish A, Thompson H, Alexander-Williams J (1978) 'Selection of patients for bile diversion surgery: use of bile acid measurements in fasting gastric aspirates.' *Gut* 19: 163–5.
21. Hoare AM, Keighley MRB, Starkey B, Alexander-Williams J (1978) 'Measurement of bile acids in fasting gastric aspirates: an objective test for bile reflux after gastric surgery.' *Gut* 19: 166–9.
22. Hyödynmaa S, Pääkkönen M, Aukee S, Korhonen K, Pääkkönen A, Länsimies E (1985) 'Quantitative radioisotope measurement of duodenogastric reflux in patients with ulcer or gastrectomized for ulcer.' *Nucl Med* 24: 107–10.
23. Karlqvist PA, Nörrby K, Svedberg J, Sjödahl R (1985) 'Enterogastric reflux surgery. A comparison between gastroduodenostomy and Roux diversion.' *Scand J Gastroenterol* 20: 861–7.
24. Kaul B, Halvorsen T, Petersen H, Grette K, Myrvold HE (1986) 'Gastroesophageal reflux disease: scintigraphic, endoscopic, and histologic considerations.' *Scand J Gastroenterol* 21: 134–8.
25. Kazem I (1972) 'A new scintigraphic technique for the study of the esophagus.' *Am J Roentgenol* 4: 681–8.
26. Klein HA, Wald A (1984) 'Computer analysis of radionuclide esophageal transit studies.' *J Nucl Med* 25: 957–64.
27. Krejs GJ, Seefeld U, Brändli HH, Bron BA, Caro G, Schmid P, Blum AL (1976) 'Gastro-oesophageal reflux disease: correlation of subjective symptoms with 7 objective oesophageal function tests.' *Act Hep Gastroent* 23: 130–40.
28. Krishnamurthy GT, Bobba VR, Langrell K (1984) 'The gallbladder emptying response to sequential exogenous and endogenous cholecystokinin.' *Nucl Med Commun* 5: 27–33.
29. Leisner B, Witte J, Kiefhaber P, Eder M, Pfeifer J, Lang G, Mayr B (1978) 'Nuklearmedizinische Diagnostik des gastroösophagealen Refluxes.' *Z. Gastroenterol* 16: 245–41.
30. Malmud LS, Fisher RS (1981) 'The evaluation of gastroesophageal reflux before and after medical therapies.' *Sem Nucl Med* 3: 205–15.
31. Martins JCR, Isaac PET, Sladen GE, Maisey MN, Edwards S (1984) 'Gastro-oesophageal reflux scintigraphy compared with pH probe monitoring.' *Nucl Med Commun* 5: 201–4.
32. Menin RA, Malmud LS, Petersen RP, Maier WP, Fisher RS (1980) 'Gastroesophageal scintigraphy to assess the severity of gastroesophageal reflux disease.' *Ann Surg* 1: 66–71.
33. Moffat RC, Berkas EM (1965) 'Bile esophagitis.' *Arch Surg* 91: 963–6.
34. Mosiman F, Sorgi M, Wolverson RL, Alexander-Williams J, Donovan IA, Harding LK (1983) 'Bile reflux after duodenal ulcer surgery: a study of 112 asymptomatic and symptom-

atic patients.' From: Abstracts 2nd International symposium on duodenogastric reflux in Switzerland, 23rd–25th of June. *Brunnen*. P63.
35. Müller-Lissner SA, Fimmel CJ, Sonnenberg A, Heinzel F, Müller R, Blum AL (1983) 'Novel approach to quantify duodenogastric reflux in healthy volunteers and patients with type I gastric ulcer.' *Gut* 24: 510–8.
36. Muhammed I, McLoughlin GP, Holt S, Taylor TV (1980) 'Non-invasive estimation of duodenogastric reflux using Technetium-99m p-butyl-iminodiacetic acid.' *Lancet* 29: 1162–5.
37. Niemela S, Heikkilä J, Lehtola J (1984) 'Duodenogastric bile reflux in patients with gastric ulcer.' *Scand J Gastroenterol* 19: 896–8.
38. Piepsz A, Georges B, Rodesch P, Cadranel S (1982) 'Gastroesophageal scintiscanning in children.' *J Nucl Med* 23: 631–2.
39. Reichelt HG, Langenberg G, Löhlein D (1977) 'Hepatobiliäre Sequenzszintigrafie in prae- und postoperativer Oberbauchdiagnostik.' *Chirurg* 48: 583–7.
40. Rhodes J, Bernardo DE, Philipps SF, Rovelstad RA, Hofmann AF (1969) 'Increased reflux of bile into the stomach in patients with gastric ulcer.' *Gastroenterology* 57: 241–5.
41. Ritchie WP (1980) 'Alkaline reflux gastritis – an objective assessment of its diagnosis and therapy.' *Ann Surg* 192,3: 288–97.
42. Rovelstad RA (1976) 'The incompetent pyloric sphincter. Bile and mucosal ulceration.' *Am J Dig Dis* 21: 165–171.
43. Safaie-Shirazzi S, DenBesten L, Zike WL (1975) 'Effect of bile salts on the ionic permeability of the esophageal mucosa and their role in the production of esophagitis.' *Gastroenterology* 68: 728–33.
44. Salo JA, Kivilaakso E (1984) 'Contribution of trypsin and cholate to the pathogenesis of experimental alkaline reflux esophagitis.' *Scand J Gastroenterol* 19: 875–81.
45. Schumpelick V (1985) 'Duodenogastraler Reflux – eine aktuelle Bilanz' *Zbl Chir* 110: 257–70.
46. Siegel JA, Wu RK, Knight LC, Zelac RE, Stern HS, Malmud LS (1983) 'Radiation dose estimates for oral agents used in upper gastrointestinal disease.' *J. Nucl Med* 24: 835–7.
47. Svedberg JB (1982) 'The bolus transport diagram: a functional display method applied to esophageal studies.' *Clin Phys Physiol Meas* 3: 267–72.
48. Thomas WEG (1984) 'The possible role of duodeno-gastric reflux in the pathogenesis of both gastric and duodenal ulcers.' *Scand J Gastroenterol* 19 (suppl 92): 151–5.
49. Tole NM (1984) 'Radiation exposure to patients during radiological examinations of the gastro-intestinal tract: intrahospital dose variations.' *Br J Radiol* 57: 297–301.
50. Tolin RD, Malmud LS, Reilley J, Fisher RS (1979) 'Esophageal scintigraphy to quantitate esophageal transit (quantitation of esophageal transit).' *Gastroenterology* 76: 1402–8.
51. Tolin RD, Malmud LS, Stelzer F, Menin R, Makler PT, Applegate G, Fisher RS (1979) 'Enterogastric reflux in normal subjects and patients with Bilroth II gastroenterostomy.' *Gastroenterology* 77: 1027–33.
52. Williams W, Krishnamurthy GT, Brar HS, Bobba VR (1984) 'Scintigraphic variations of normal biliary physiology.' *J Nucl Med* 25: 160–5.
53. Wolverson RL, Sorgi M, Mosiman F, Donovan IA, Harding LK, Alexander-Williams J (1984) 'The incidence of duodenogastric reflux in peptic ulcer disease.' *Scand J Gastroenterol* 19 (suppl 92): 149–50.

11. Nuclear medicine in inflammatory bowel diseases

ANDREAS L. HOTZE and HANS J. BIERSACK

Introduction

Nearly 13 years ago inflammation imaging using labeled white blood cells (WBCs) has been introduced in clinical nuclear medicine.

In 1977 Thakur et al. [1] first described a method for labeling leukocytes with ^{111}Indium oxine.

For the diagnosis of inflammatory abdominal diseases this method has been used by many nuclear medicine physicians, who all reported a high sensitivity and specificity (synopsis see: [2]).

The accuracy of the method is based even on the fact that excepted in liver and spleen normally no ^{111}In labeled WBCs are present within the abdominal area, and especially not within the bowel lumen. The presence of In WBCs in the abdomen may thus be interpreted as a strong evidence for infection. On the other hand, one must be aware of that in the recent studies the respective disease had a high prevalence,and this has been taken into statistical evaluation, and it must be critically mentioned in the retrospetive consideration of those data.

For the last ten years ^{111}In oxine has been accepted as the radiopharmaceutical of first choice for cell labeling [18, 19]. Attempts of labeling leukocytes or other cells with pharmaceuticals attached with ^{99m}Tc failed to be successful. In 1986, Peters et al. [3] and Joseph et al. [5] first described a method for cell labeling with ^{99m}Tc-HMPAO.

Also in 1986, Schuemichen [4] first described imaging inflammatory bowel disease using ^{99m}Tc-HMPAO as labeling agent.

HMPAO has been originally developed for brain SPECT studies, e.g. to measure regional cerebral blood flow.

We also have used HMPAO leukocyte imaging for the detection of infections outside the abdomen. [7–9] Although the use of this technique for diagnosing abdominal infection has not yet been pursued.

We continue to use ^{111}In oxine leukocytes for detection of abdominal lesions since free ^{99m}Tc appears in the gall bladder and in the biliary tree, and then is transported to more distal bowel sections after injection of

H.J. Biersack and P.H. Cox (eds), Nuclear Medicine in Gasteroenterology, 169–176

HMPAO labeled leukocytes. This free activity is unfavorable, and renders the diagnosis of minimal bowel infection more difficult.

In 80% of patients referred with suspected osteomyelitis we could demonstrate this unspecific abdominal activity. The biliary excretion of free ^{99m}Tc results in a positive visualization of parts of the small bowel.

We thus continue to use and also recommend cell labeling with ^{111}In oxine for imaging inflammatory lesions of the abdomen.

Methods

Cell separation

There are a variety of techniques for labeling cells which in general involve a series of sedimentations, centrifugation, and washing steps to separate leukocytes from the plasma. Our protocol involves the isolation of mixed leukocytes from 45 ml venous blood which has been anticoagulated with 2500 IU of heparin. After the blood is withdrawn, hydroxyethyl starch (5 ml) is added to the syringe followed by a 30 to 40 min sedimentation. The supernatant is then removed, resuspended, and centrifugated at 70–90 × g. The cell pellet consists of erythrocytes and granulocytes and is then again resuspended in NaCl and centrifugated.

Cell labeling

The cell pellet with the remaining blood compounds is then incubated for 15 min with 18.5 MBq ^{111}In oxine (or with 500 MBq ^{99m}Tc HMPAO in suspected osteomyelitis) and resuspended with 10 ml NaCl before injection. We use freshly delivered ^{111}In oxine and one step labeled HMPAO, respectively.

Pure granulocyte preparation

For further differentiation white blood cells can be separated by density gradient centrifugation, e.g. Ficoll Hypaque or Percoll. Both procedures require new centrifugation and washing steps. In our laboratory the complete separation and labeling procedure requires between 2 to 2.5 h.

Anti granulocyte antibodies

For clinical studies we have used a new *monoclonal* antibody which is directed against the granulocyte epitope BW 250/183. Our initial studies have focussed on patients with osteomyelitis of the lower extremities. We obtained excellent results with this new antibody and have performed whole body scans in all patients investigated. Nonspecific accumulation of activity in the abdomen was not observed in any case, and only very low accumulation in the liver and spleen, which would allow clear visualization of inflammatory abdominal lesions.

Quality controls

Cell contamination. In both labeling procedures a considerable number of erythrocytes and platelets which will also be labeled has to be taken into account.

Generally, the highest cell contamination is found in mixed leukocyte preparations. Cell numbers of 10 to $15 \times 10^{12}\,l^{-1}$ leukocytes, and 80 to $100 \times 10^{9}\,l^{-1}$ platelets were found. Also, the ratio between granulocytes and lymphocytes is more unfavorable when using mixed cell preparation. The quantitative evaluation of cell contamination is not, however, an essential part of routine quality controls.

Viability controls. Trypane blue or eosine red tests are easy to perform for evaluation of cell viability and these staining procedures becoming an important aspect of these labeling procedures. 100 μl of cell suspension are diluted with one droplet of the respective stain solution, and the stained cells were then microscopically counted after a 30 min incubation. The ratio between stained and unstained cells represents an index of the number of cells damaged during the separation and radiolabeling procedure.

With all cell isolation methods the number of avital cells in relation to total cells is usually less than 5%. It is our opinion that viability tests should be performed in all cell radiolabeling procedures.

Labeling yield. The ratio of cell bound radioactivity to the total activity is calculated by the following method. An aliquot of the cell suspension after labeling is centrifugated, the supernatant is removed and both the pellet and the supernatant then counted. The radiolabeling yield is the ratio between pellet activity to total activity.

$$\mathrm{Ly} = \frac{\mathrm{Ap}}{\mathrm{Ap} + \mathrm{As}} \qquad (1)$$

Ly is the labeling yield; Ap is the activity in pellet; As is the activity in supernatant

When using the ^{111}In oxine radiopharmaceutical, the labeling efficiency is usualy 90% or greater. The yield decreases to 30 to 60% when ^{99m}Tc HMPAO is used as the radiolabeling agent.

Scanning

Localization of inflammatory lesions. To localize an inflammatory lesion we obtain planar scintigrams of the total abdomen 4 and 24 h p.i. using a LFOV camera fitted with a high energy collimator. There are efforts, however, to obtain very early images (e.g., 30 min p.i.), and include these as part of the widely used standard protocol. We feel that early images within 1 h p.i. could lead to false positive findings, e.g., in hyperemia, since the blood clearance of labeled cells is not yet complete at this early time interval. Our own experiences have shown that it is not necessary to strictly maintain the time of 4 h p.i., but also periods between 2 and 6 h can be chosen without loosing any image qualitiy or information.

Estimation of inflammatory activity. There are essentially two possibilities for quantification of inflammatory lesions

a) Determination of fecal ^{111}In excretion [2, 26]. This procedure requires a complete stool collection from the injection time until 4 days p.i. The excreted activity is then measured and related to the total injected activity. Assuming the stool collection is complete, this method may be objective and very accurate, but is time consuming and impractical.
b) The second method consists of a quantitative comparison of cell uptake in a lesion (when present) and a bone marrow region by means of ROI technique. This method provides the advantage that the estimation of disease activity is available within a few hours, approx. 4 h p.i. It should be stressed, however, that the method a) is suitable only in in-patients since a complete stool collection must be essentially guaranteed.

   Method b), however, can be employed on both an in- and out-patients basis.

Differential diagnosis. Leukocyte imaging can detect inflammatory abdominal lesions with a high sensitivity. Due to the different accumulation patterns, it is also possible to differentiate between chronical bowel infection, fistulas, and abscesses.

a) Chronical bowel diseases are positive on early scans, and on the delayed scan more distal bowel content becomes positive due to aboral intraluminal cell transit.
b) Abscesses are clearly visible as accumulation on early scan. The delayed scan reveals an unchanged positive contrast which is due to the fact that the cells maintain in the abscess.
c) Fistulas become faintly positive in the early imaging, but they normally empty their content into the bowel lumen so that the delayed images are showing a positive bowel content, and the images are comparable to those which are obtained in chronical bowel infection.

Results

According to published data and to our own results, one can definitely conclude that an inflammatory bowel lesion will be detectable by ^{111}In oxine WBCs with a high accuracy. Comparative studies (^{111}In oxine leukocytes and endoscopy) showed a significant correlation ($r = 0.8$) between both methods with respect to localization of inflammation. Low grade inflammatory lesions, however, may fail to be detected by scintigraphy. In our cases we found a few false negative scans in patients with relapse of Crohn's disease with very low inflammatory activity which was also proven histologically. Also, pretreated inflammation particulary in partial remission may not be detected on inflammation scan. Typically, chronic bowel inflammation will be visualized positive on early scans (2–6 h p.i.), and also be correctly localized. On the delayed scans the activity will normally be found more distal due to intraluminal cell transportation, and more distal bowel areas will be visualized. Abscesses also show a typical accumulation pattern, since the labeled cells rapidly migrate to these sites with clear visualization of the infection. Cells are maintained, however, for a longer period at the site of infection causing unchanged findings on delayed images with still positive abscess and only weak bowel content visualization. These different accumulation pattern can be taken for differential diagnosis of the latter. Fistulas behave very similar to chronic bowel disease. They are positive on early scans but then, after emptying their content into the bowel lumen, more distal bowel segments will be visualized positive on the delayed images in the same manner as chronic bowel infection.

Estimation of activity of inflammatory bowel disease

a) Determination of fecal ^{111}In excretion. Using the fecal excretion of ^{111}In over a 4 days period, the severity of inflammation can be objectively assessed when the excreted activity (correlated to the number of WBCs) is related to the injected total activity. In normal adult volunteers this ratio is essentially less than 2% . Also, a good correlation of this parameter with the BSR and the activity index as suggested by van Hees [10] is reported. However, it must be stressed again that fecal cell secretion measurement can be only performed with a reliable accuracy in in-patients since a complete stool collection over 4 days is necessary. An incomplete stool collection would lead to false negative results.
b) In contrast to the above procedure, the activity estimation of inflammation by comparison of cell uptake of an inflammatory bowel lesion and a reference region (e.g., bone marrow, soft tissue) does not depend upon the cooperation of the patient and the technologists. Furthermore, the results are available immediately after the investigation.

Finally, the limitations of the nuclear medicine technique to localize abdominal inflammation and estimate its severity are as follows:

1. *Specificity*. The specificity is much lower than the sensitivity. Leukocyte imaging can only provide the detection of an inflammatory lesion but cannot discover its origin. In consequence, this method cannot differ-

entiate, for instance, between Crohn's disease, ulcerative colitis, and an yersinia enteritis [20, 23, 25, 11].

2. *False positive findings*. False positive results can be obtained in gastrointestinal bleeding, multiple barium enemas, and bowel infarction. Also, false positive abdominal findings are observed due to swallowed leukocytes in inflammatory diseases of the respiratory tract, e.g., in cystic fibrosis, pneumonia, sinusitis, esophagitis, and parotitis [21, 15].
3. *False negative findings*. False negative findings are primarily expected in very small inflammatory changes. Moreover, one has to consider false negative scans in initially treated disease when the inflammation is partially in remission. A further possibility may also result from an incorrect labeling, e.g., too many nonvital cells [16, 17, 24].

More recently, false negative as well as false positive findings have also been reported in lesions outside the abdomen [12–14].

Clinical applications

Chronical inflammatory bowel disease

Initial scintigraphic diagnosis is only indicated and may often be required when there is danger of bowel perforation, and either a partial endoscopy can be performed, or an endoscopy is impossible.

In other patients leukocyte imaging can be employed supplementary to endoscopic or radiological investigations with respect to localization and activity estimation.

It is also possible to employ leukocyte imaging as an alternative method in the follow up for activity estimation, or complementary to endoscopy and biopsy.

Abscesses/fistulas

Generally, abscesses are easily detectable by this method and can be differentiated from chronic bowel disease and fistulas due to their accumulation pattern.

The question whether leukocyte imaging is the principal diagnostic method depends on how fast the clinical diagnosis is required, and the general condition of the patient.

In non-acute patients, where the diagnosis must not be known immediately, leukocyte imaging can be the first diagnostic tool. In case of positive findings, other modalities can be used specifically. A negative inflammation scan excludes a floride infection with a high probability.

In seriously ill patients in whom an immediate therapeutic consequence is required other modalities should be first employed. However, most patients are usually referred for leukocyte imaging when TCT and sonography results show normal findings.

Differential diagnosis of chronical bowel diseases fistulas/abscesses
Due to the different accumulation patterns, the above diseases are easily differentiated by means of leukocyte imaging when using early and delayed scans.

References

1. Thakur ML, Lavender JP, Arnold RW et al. (1977) '^{111}In labelled autologous leukocytes in man.' *J Nucl Med* 18: 1014.
2. Becker W, Fischbach, W, Reiners Chr et al. (1987) 'Indium-Oxin markierte Granulozyten bei chronisch entzuendlichen Darmerkrankungen und abdominellen Abszessen – Methodik, Ergebnisse, Indikationen.' *Nuklearmediziner* 10: 127.
3. Peters AM, Danpure, HJ, Osman S et al. (1986) 'Clinical experience with ^{99m}Tc-Hexa-methyl-propyleneamineoxime for labeling leukocytes and imaging inflammation.' *Lancet* II: 946.
4. Schuemichen C, Schoelmerich J (1986) '^{99m}Tc-HMPAO labelling of leukocytes for detection of inflammatory bowel diseases.' *Nuc Compact* 17: 274.
5. Joseph K, Damman V, Engeroff G, Gruner KR (1986) 'Markierung von Leukoyten mit ^{99m}Tc-HMPAO: Erste klinische Ergebnisse.' *Nuc Compact* 17: 277.
6. Saverymuttu SH, Camilieri M, Rees M et al. (1986) '^{111}In granulocyte scanning in the assessment of disease extent and disease activity in inflammatory bowel disease.' *Gastroenterology* 90: 1121.
7. Hotze A, Briele B, Wolfe F et al. (1988) 'Localization and activity of inflammatory bowel disease using ^{111}In leukocyte imaging.' *Nucl Med* 27: 83.
8. Hotze, A (1988) 'Leukozyten-Szintigraphie in der gastroenterologischen Nuklearmedizin.' *Nuklearmediziner* 11: 347.
9. Hotze A, Briele B, Wolf F, Biersack HJ (1987) 'Estimation of extent and activity of inflammatory bowel diseases using ^{111}In oxine labeled leukocytes.' *J Nucl Med* 28: 659.
10. van Hees PAM, van Elteren PA et al. (1980) 'An index of inflammatory activity in patients with Crohn's disease.' *GUT* 21: 279.
11. Haentjens M, Piepsz A, Shell-Frederick E et al. (1987) 'Limitations in the use of ^{111}In oxine labeled leukocytes for the diagnosis of occult infection in children.' *Pediatr Radiol* 17: 139.
12. Kim EE, Pjura Ga, Lowry PA et al. (1987) 'Osteomyelitis complicating fracture: Pitfalls of ^{111}In-leukocyte scintigraphy- *AJR* 148: 927.
13. Wuckich DK, Abren SH, Callaghan JJ (1987) 'Diagnosis of infection by preperative scintigraphy with indium-labelled white blood cells.' *J Bone Joint Surg* 69A: 1353.
14. van Nostrand D, Abreu SH, Callaghan JJ et al. (1988) '^{111}In-labeled white blood cell uptake in non-infected closed fracture in humans: Prospective study.' *Radiology* 167: 497.
15. Syrjala MT, Valtonen V, Liewendahl K et al. (1987) 'Diagnostic significance of ^{111}In granulocyte scintigraphy in febrile patients,' *J Nucl Med* 28: 155.
16. Becker, W, Boerner W (1988) '^{99m}Tc-HMPAO zur Markierung von Leukozyten und Thrombozyten. Besonderheiten und diagnostische Konsequenzen.' *Fortschr Roentgenstr* 148: 69.
17. Danpure HJ, Osman S (1988) 'Optimum conditions for radiolabelling human granulocytes and mixed leucocytes with ^{111}In-tropolonate.' *Eur J Nucl Med* 14: 537.
18. McAfee JG, Subramanian G, Gagne G et al. (1987) '^{99m}Tc-HMPAO for leukocyte labeling – experimental comparison with ^{111}In-oxine in dogs.' *Eur J Nucl Med* 13: 353.
19. Arndt JW, Crama-Bohbouth GE, Verspaget HW et al. (1989) 'Image quality and radiopharmaceutical parameters of ^{111}In granulocytes in scintigraphy of inflammatory disease.' *Eur J Nucl Med* 15: 197.

20. Crama-Bohbouth GE, Arndt JW, Pena AS et al. (1988) 'Value of ^{111}In granulocyte scintigraphy in the assessment of Crohn's disease of the small intestine.' *Digestion* 40: 227.
21. Seybold K, Locher TJ, Coosemans C et al. (1988) 'Immunoscintigraphic localization of inflammatory lesions: clinical experience.' *Eur J Nucl Med* 14: 587.
22. Ybern A, Martin-Comin J, Gine JJ et al. (1986) '^{111}In-oxine-labeled autologous leucocytes in inflammatory bowel disease: New scintigraphic activity index.' *Eur J Nucl Med* 12: 341.
23. Poitras P, Carrier L, Chartrand R et al. (1987) '^{111}In leukocyte scanning of the abdomen. Analysis of its value for diagnosis and management of inflammatory bowel disease.' *J Clin Gastroenterol* 9: 418.
24. Butler JA, Marcus CS, Hennemann PL et al. (1987) 'Evaluation of ^{99m}Tc leukocyte scan in the diagnosis of acute appendicitis.' *J Surg Res* 42: 575.
25. Froehlich JW, Field SA (1988) 'The role of ^{111}In white blood cells in inflammatory bowel disease.' *Sem Nucl Med* 18: 300.
26. Becker W, Fischbach W (1988) 'Aussagemoeglichkeiten von ^{99m}Tc-HMPAO- und ^{111}In-Oxin markierten Granulozyten bei chronisch entzuendlichen Darmerkrankungen.' *Nuklearmediziner* 11: 89.

12. Detection and localization of gastrointestinal bleeding sites with scintigraphic techniques

ALAN SIEGEL and ABASS ALAVI

Introduction

Gastrointestinal (GI) hemorrhage is a serious medical problem with a mortality rate of approximately 10% for major bleeding episodes. About 15% of bleeds will not stop spontaneously and early diagnosis and treatment are essential in decreasing morbidity and mortality. Prognosis is related to the following factors [1]:

1. Age of the patient.
2. Rapidity of the bleed.
3. Single versus recurrent episodes.
4. Concomittant illness.
5. Rapidity of surgical intervention.

Diverticulosis and arteriovenous malformations (AVMs) are among the most common causes of lower GI bleeds. Diverticular bleeds are more frequent in patients over 50 years of age and tend to occur in the right side of the colon. AVMs bleed more frequently than diverticula, but the hemorrhage is usually less severe. These lesions include angiodysplasias and telangiectasias and are also more common in the elderly. Polyps, Meckel's diverticula and inflammatory bowel disease should be considered in younger patients.

Upper gastrointestinal bleeding is more common than lower. The more frequent causes include peptic ulcer disease, gastritis and esophageal varices.

Localization of bleeding is essential for proper management. History and physical examination alone may not even be sufficient to determine if bleeding is occuring from the upper or lower GI tract. Several techniques for diagnosis are available. Fiberoptic endoscopy is a useful method, detecting the bleeding site in the GI tract in about 90% of cases [1]. Therapy of upper gastrointestinal bleeding may be performed with endoscopy, including sclerotherapy and electrocoagulation. Endoscopy, however, may be limited by the presence of an excessive amount of blood in the lumen of, for example, the large bowel, obscuring visualization. Also, most of the small bowel is not easily accessible to the endoscope.

H.J. Biersack and P.H. Cox (eds), Nuclear Medicine in Gasteroenterology, 177–190

Angiography is an invasive procedure with inherent risks and a reported sensitivity of 66% with bleeding rates of greater than 0.5 ml/min [1]. Detection is poor at lower bleeding rates. Angiography is also useful as a therapeutic technique. Vasopressin infusion and transcatheter embolization through a selectively placed catheter are effective methods for the treatment of bleeding [2, 3].

Nuclear scintigraphy for the detection of gastrointestinal bleeding (mostly lower GI bleeding) includes several techniques which are rapid, easily performed, of minimal risk and relatively low radiation exposure to the patient and have a high sensitivity. These examinations are excellent as initial procedures for the localization of gastrointestinal bleeding.

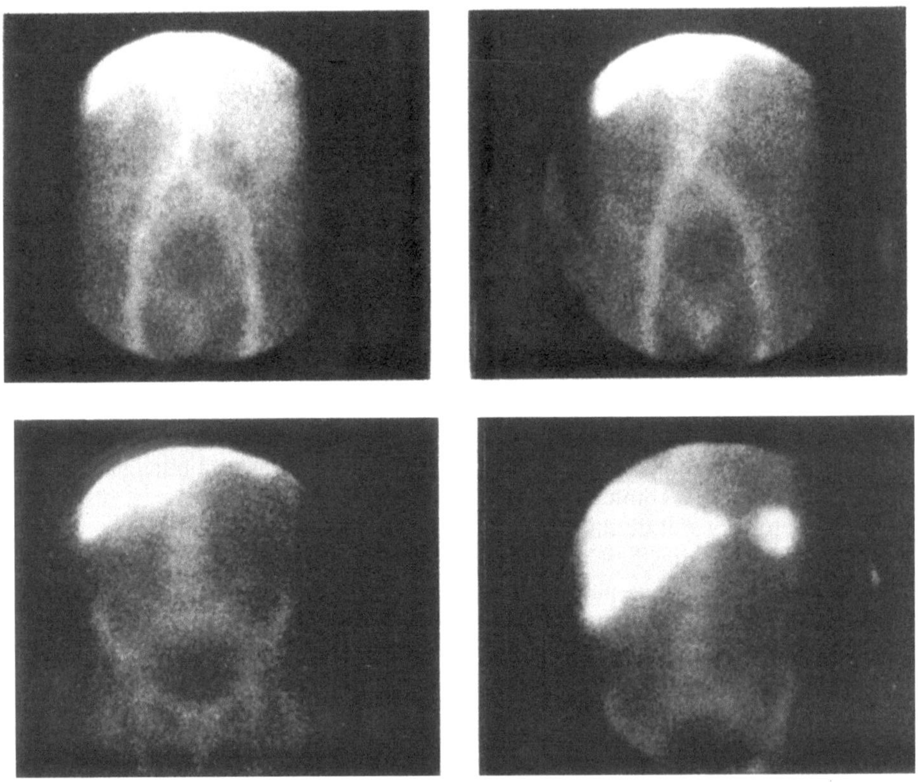

Figure 1. Normal Tc–SC scan. Images from a normal Tc–SC study show the usual liver, spleen and bone marrow activity at 5, 15, 30 and 45 minutes after injection. Vascular structures are visualized early but are not seen after 15 min. (Reprinted with permission from Alavi A, Ring EJ (1981). AJR 137: 741–748).

Techniques

^{99m}Tc sulfur colloid

After intravenous administration, ^{99m}Tc sulfur colloid is phagocytozed by the cells of the reticuloendothelial system in the liver, spleen and bone marrow The half time of the agent in the intravascular space is approximately 2.5 min. If GI bleeding is actively occuring at the time of injection, some of the radionuclide will be extravasated into the bowel. After approximately 12 to 15 min, little of the agent will remain within the circulatory system. This will allow for a good target to background ratio but, for the same reasons, no additional activity will extravasate once the radiopharmaceutical is cleared from the circulation (Fig. 1).

Ten to fifteen millicuries of ^{99m}Tc SC are administered intravenously after the patient is placed in the supine position with the camera anterior to the

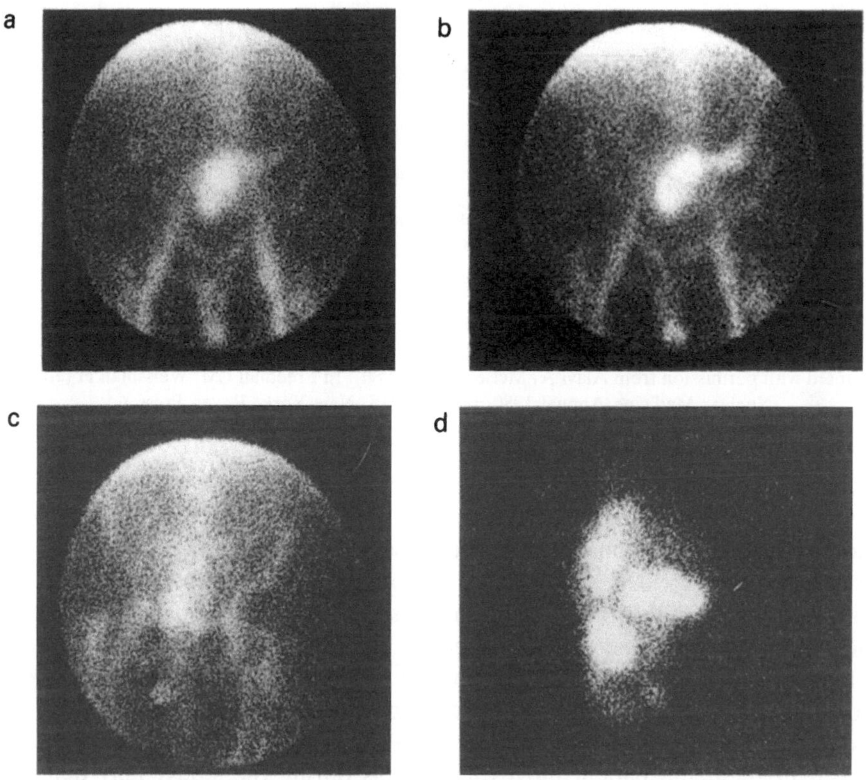

Figure 2. Bleeding site in the distal descending colon: Tc–SC. Activity is initially seen in the distal descending colon and travels retrograde and anterograde (Fig. 2a–c). The patient had a bowel movement during the study and activity was present in the bedpan (Fig. 2d) (Reprinted with permission from Alavi A (1982). Semin Nucl Med 12: 126–138).

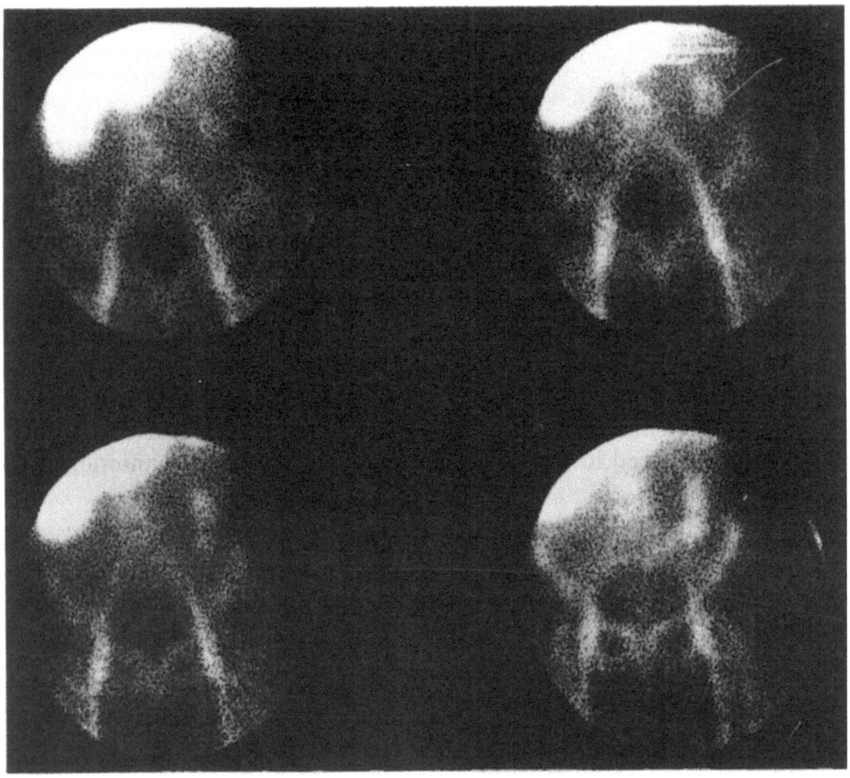

Figure 3. Bleeding site in the splenic flexure: Tc–SC. A focus of activity is present in the splenic flexure and is seen to progress down the descending colon to the sigmoid. Arteriography was unsuccessful in this case because of complete occlusion of the inferior mesenteric artery (Reprinted with permission from Alavi A, McLean GK (1980). In Freeman LM, Weissman H (eds), Nuclear Medicine Annual 1980. pp. 177–218. New York: Raven Press.).

abdomen [4]. The camera should be positioned so that the lower edge of the liver is in the uppermost portion of the image. The intensity should be adjusted so that the bone marrow can be visualized.

Consecutive 750000 count images are obtained for 15 min. If a bleeding site is localized at this time, further images are taken until movement of the activity is noted so that the exact sight of hemorrhage is determined. If no bleeding site is seen, an LAO view of the upper abdomen may be obtained for better visualization of the splenic flexure and proximal small bowel. In addition, imaging of the glove following a rectal examination or of stool in the bedpan, if the patient moves his bowels during the study, may detect activity from a bleeding site in the rectum which may occur with an otherwise negative study (Fig. 2).

Movement of activity within the bowel lumen, either antegrade or retrograde, is necessary for confirmation of bleeding (Figs 3, 4). An area of

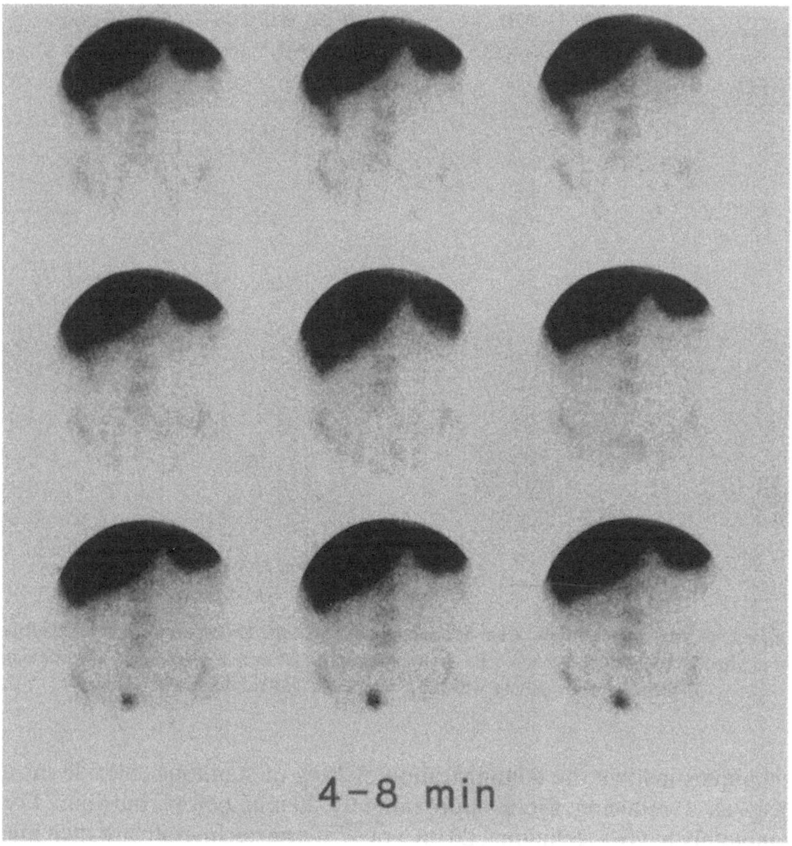

Figure 4. Bleeding site in the hepatic flexure: Tc–SC. The earliest images demonstrate the bleeding site in the region of the hepatic flexure, just below the right lobe of the liver. Several minutes later, the activity has moved through the colon and is seen pooling in the recto-sigmoid area. This case demonstrates how delayed imaging may be misleading in some patients. This patient had bleeding from severe diverticulosis of the ascending colon and underwent right hemicolectomy.

activity which does not move is unlikely to represent bleeding. Other causes, such as an accessory spleen, should be considered.

Labeled red blood cells

The labeling of red blood cells with ^{99m}Tc pertechnetate allows for visualization of the intravascular space for up to several hours (Fig. 5). The extravasation of labeled red blood cells into the bowel may therefore be imaged as well. Increasing intensity of activity along the course of the bowel, antegrade or retrograde, will confirm the presence of hemorrhage (Fig. 6a–c).

Red blood cells can be labeled by several different techniques. In vivo

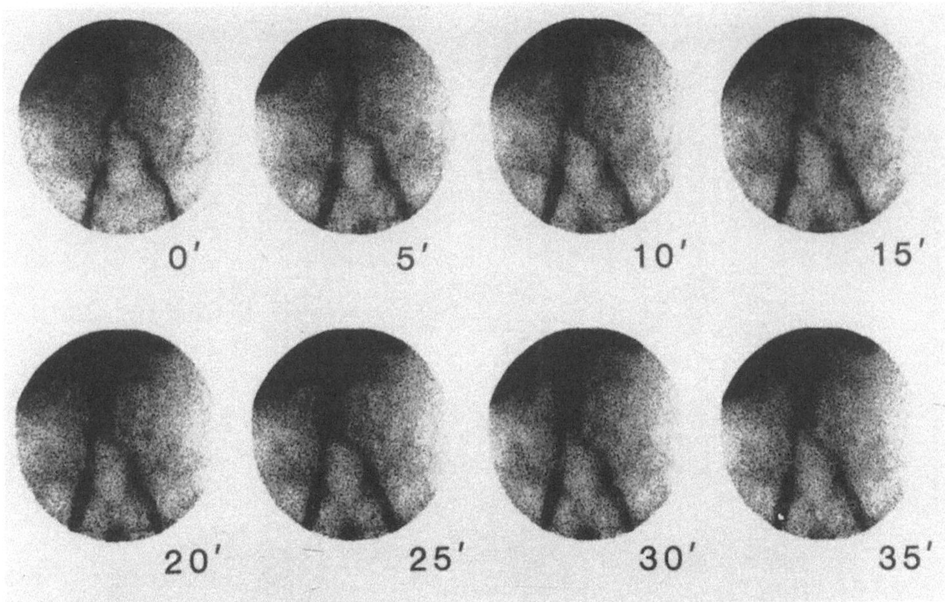

Figure 5. Normal labeled RBC scan. Sequential images of the abdomen were obtained following the in vitro labeling of red blood cells. Images taken for 35 min are displayed and show normal intravascular activity with no evidence of gastrointestinal bleeding.

techniques involve the administration of 1 mg of stannous chloride intravenously as a reducing agent approximately 20 min before labeling. For the completely in vivo technique, 20 mCi of ^{99m}Tc pertechnetate are then injected intravenously into the patient. This technique is simple and fast but results in the poorest labeling efficiency, with only about 70% of the tracer coupled to RBCs (Fig. 7) [5].

The in vivtro method (a combination of the in vivo and in vitro approaches) is a modification of the method described above. Here 3 ml of blood are withdrawn (15 min following the administration of 1 mg of stannous chloride IV) into a syringe containing 20 mCi of ^{99m}Tc pertechnetate and heparin. The syringe is then shaken every minute for 10 min and the blood is then reinjected into the patient [6]. There is a somewhat superior labeling efficiency using this method, but a fair amount of unlabeled activity is still seen in the images.

The most appropriate labeling technique for the detection of bleeding is the in vitro method. Here the patient's RBCs are withdrawn into a syringe containing 20 mCi of ^{99m}Tc pertechnetate, heparin and stannous citrate. After 5 min, the RBCs are washed with saline and readministered to the patient. Although this is the most time consuming of the different techniques, it allows for a 95% labeling efficiency.

To image with labeled red blood cells, the patient is positioned supine with the camera anterior to the abdomen. The field of view should extend

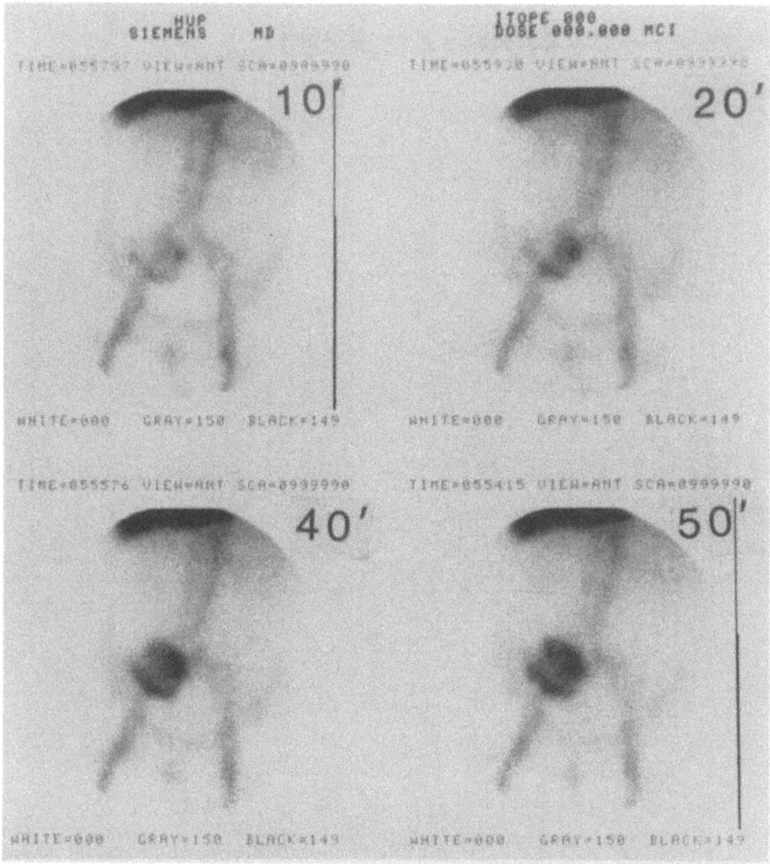

Figure 6. Bleeding in the cecum: Tc–RBCs. Activity extravasates from a bleeding site in the cecum. The presence of vascular structures clearly identifies this as a labeled RBC study (Fig. 6a). An angiogram was performed on the same patient via a catheter in the superior mesenteric artery (Fig. 6b). The image from the late phase of the study shows extravasation of contrast in the cecum (Fig. 6c).

from the inferior aspect of the heart to the inferior pelvis. Images are obtained using a low energy parallel hole collimator with a large field of view for 45 min in much the same way as described or ^{99m}Tc sulfur colloid [7]. The intensity of the images should be set so that the aorta, inferior vena cava and iliac vessels are visualized. Unlike sulfur colloid, delayed images can be obtained for up to 24 h after injection.

Discussion

Most episodes of GI bleeding are intermittent and a majority of patients will have stopped bleeding before diagnostic procedures can be completed. It

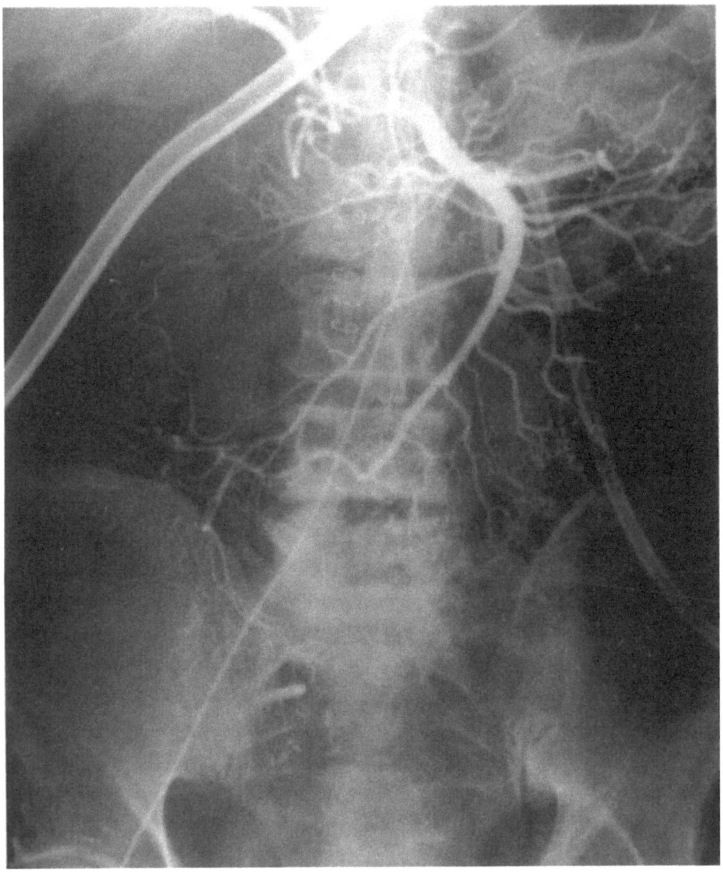

Figure 6b.

is therefore beneficial to institute the work-up as soon as possible after presentation.

Active lower GI bleeding will usually present as hematochezia although rapid upper GI bleeding may do so as well. Upper GI bleeding will more frequently present as melena or hematemesis. The initial step in the work-up of GI hemorrhage, following history and physical, should be placement of a nasogastric tube. If blood is present in the gastric contents, an upper GI bleed must be considered and upper endoscopy frequently follows. If the gastric contents contain no blood, a lower GI bleed may be present. Unless the patient is hemodynamically unstable and emergent surgery is required, a nuclear GI bleeding scan should be the next test.

Most patients will no longer be bleeding by the time they are scanned. Only about 15–20% of immediately performed scans will show active bleeding [8]. Therefore, a patient should be scanned as soon as possible following an episode of hematochezia.

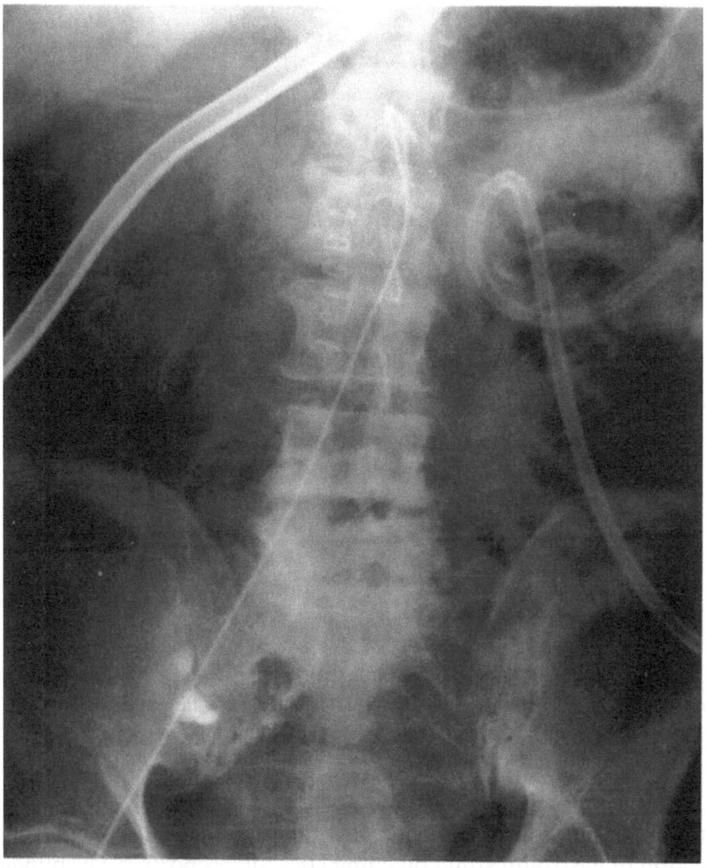

Figure 6c.

While bleeding scans are quite useful as screening tests for lower GI bleeds, the routine use for upper GI bleeds is not justified [6, 9–12]. Intense liver activity on Tc–SC scans obscures much of the upper abdomen. Using labeled RBCs, high background activity in the upper abdomen decreases sensitivity. Also, cardiac activity may obscure the lower esophagus and free technetium can cause false positives in the stomach. There are, however, instances in which upper GI bleeding sites have been detected during scanning with both Tc–SC and labeled RBCs (Figs 8, 9).

Tc–SC as an agent for the detection of gastrointestinal hemorrhage was proposed and successfully utilized in animals in 1972 [13]. The data from experiments in dogs showing successful detection of bleeding sites was published in 1977 [14]. The use of radiolabeled intravascular markers for the detection of GI bleeding was first proposed in 1977 [15]. ^{99m}Tc labeled albumin was the first agent used but has since been replaced by labeled RBCs which allows for superior image quality and fewer artifacts [16, 17].

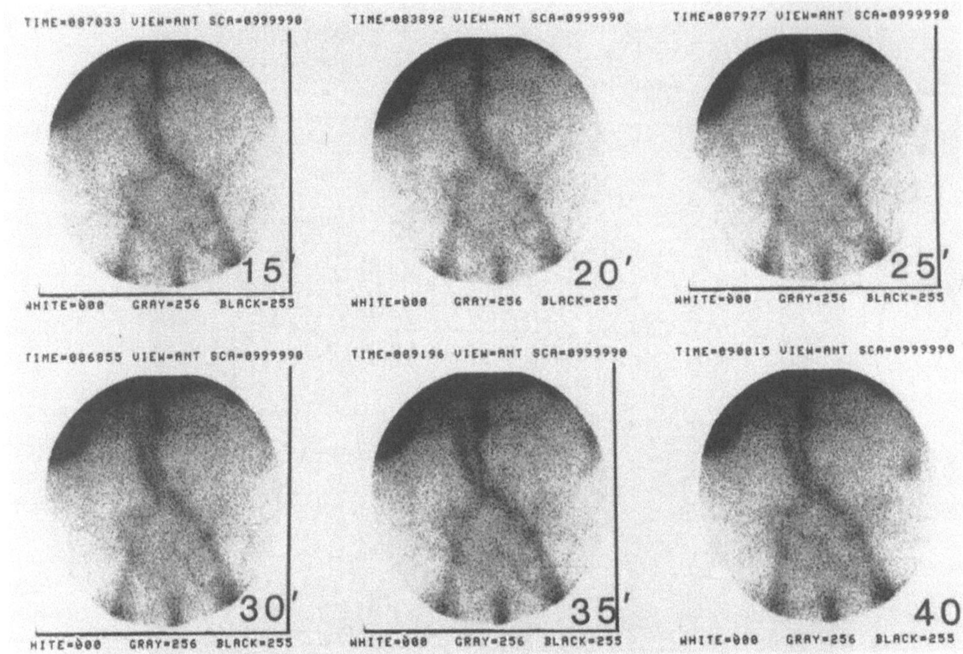

Figure 7. Bleeding in the descending colon: Tc–RBCs. A bleeding site is present in the upper descending colon and is best seen on the later images. Low target-to-background ratio is probably due to significant blood pool activity in the abdomen. Colonoscopy revealed severe diverticulosis of the descending colon.

Scintigraphic methods have been favorably compared to angiography in the detection of gastrointestinal bleeding [18–20]. Reports have shown ^{99m}Tc sulfur colloid bleeding scans to be more sensitive for the detection of bleeding sites than angiography and to have few false negatives. The bleeding scan has been shown to detect bleeding rates as low as 0.05 to 0.10 ml/min. These rates are 5–10 times lower than can be seen with arteriography. In addition, bleeding scans are much more successful at detecting venous bleeds [21].

It has been argued that scanning with labeled RBCs is more useful than sulfur colloid because it allows for reimaging if an episode of active bleeding should occur, even up to hours after the injection of the radiotracer. Patients who re-bleed after leaving the department may be returned for additional images. It must be remembered that the scan is often only useful if it can localize the site of bleeding. If images are obtained after bleeding has started, the activity may have already moved within the bowel making detection of the bleeding site impossible (Fig. 4). Bowel transit may be markedly increased in the presence of blood. Scanning with labeled RBCs is most advantageous when continuous images may be taken for extended periods of time. In addition, if Tc–SC is used, the patient may be reinjected as soon as another bleeding episode is suspected, even if it is only minutes later. The increased

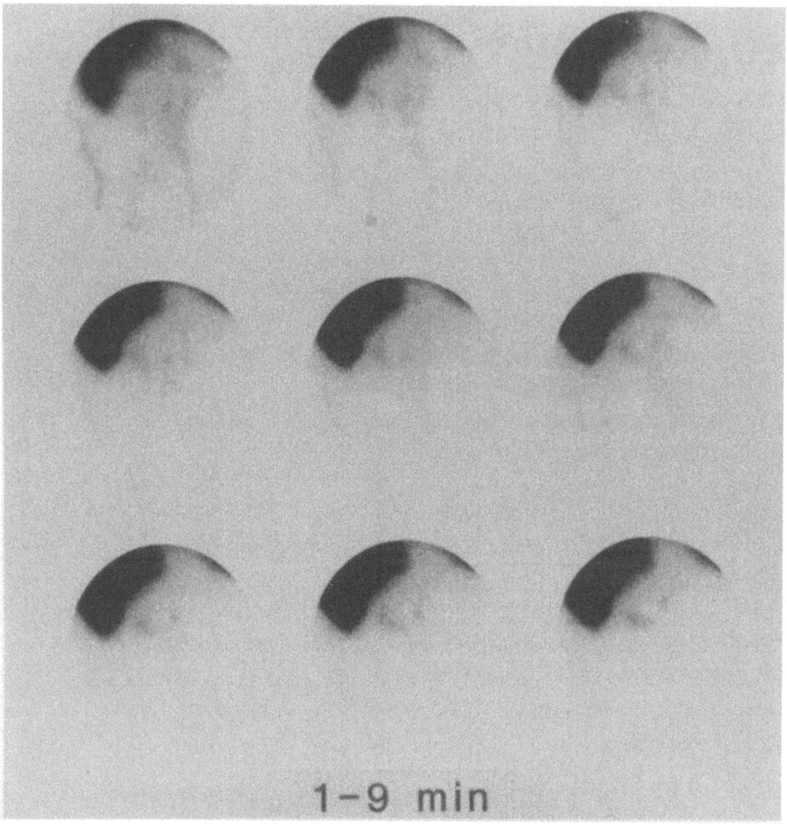

Figure 8. Bleeding site in the duodenal bulb: Tc–SC. Sequential images over the first 9 min of the study demonstrate activity in the sweep of the duodenum.

radiation exposure may be justified since its concomittant risk is lower than that of repeated transfusions or blind surgery, especially in the elderly patient.

The sensitivity of detection of GI bleeding using ^{99m}Tc labeled RBCs is somewhat lower than that of ^{99m}Tc sulfur colloid using the in vivo and in vivtro methods [22]. The improved target-to-background ratio of the in vitro method makes it the method of choice for labeling red blood cells for the detection of bleeding sites. Labeling techniques, especially the in vivo method, result in some amount of free ^{99m}Tc pertechnetate. Therefore, activity may also be seen in kidneys, bladder and stomach. Activity excreted by the stomach can mimic blood in the small and large bowel. Another difficulty with labeled RBCs is with detection of areas of increased vascularity. AVMs and varices may show intense activity and can prove confusing.

If it is necessary to take a patient to surgery, the bleeding scan, when positive, can be an important guide to determine the portion of bowel to be

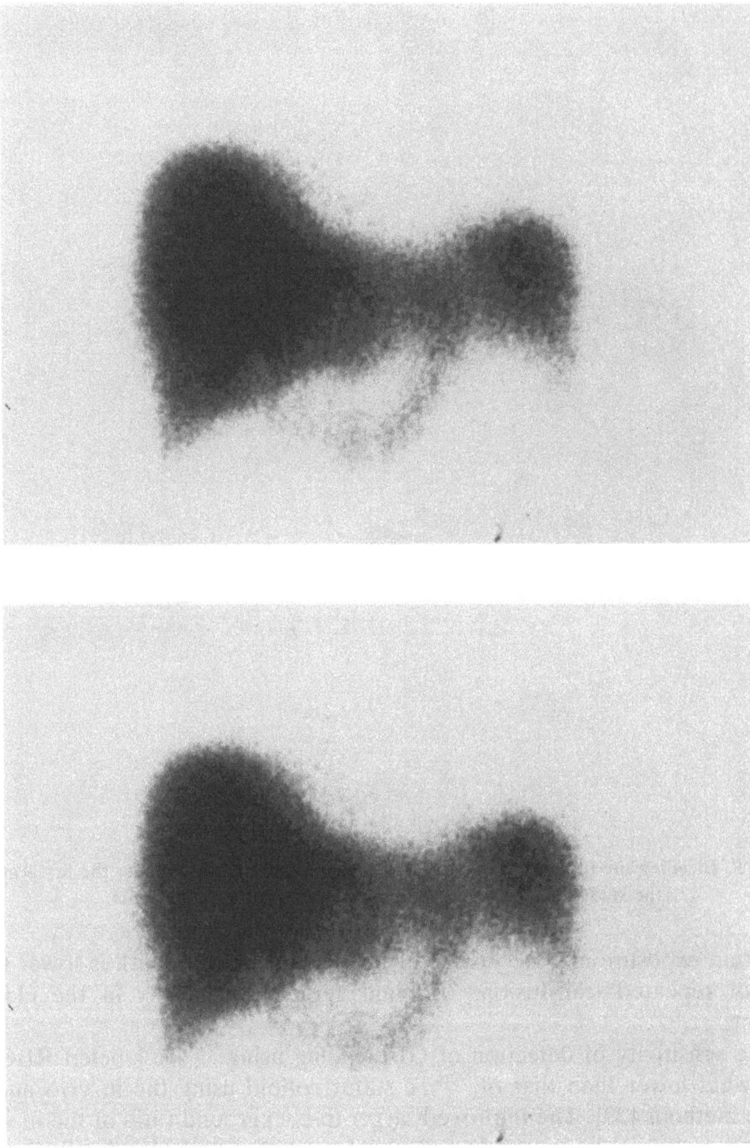

Figure 9. Gastric hemorrhage: Tc–Sc. Although limited, the Tc–SC may, at times, demonstrate UGI bleeds. Figures 9a and 9b clearly demonstrate activity in the stomach.

resected. In addition, nuclear scans can be used as a guide for the non-surgical management of bleeding which may be especially beneficial for patients who are poor surgical candidates. The scan may guide angiographers in the use of embolic therapy, using materials such as coils, Gelfoam and Ivalon or for vasopressin infusion [23–26]. In addition, the scan may be an aid to endoscopists using electrocoagulation [27, 28]. Finally, nuclear scanning is an excellent method of follow-up after invasive procedures have been used to stop bleeding.

References

1. Sullivan BH Jr. 'Gastrointestinal bleeding' In Farmer RG, Achkar E, Flechler B (eds) (1983) *Clinical Gastroenterology*. New York: Raven Press, pp 17–21.
2. Rosch J, Dotter CT, Brown MJ (1972) 'Selective arterial embolization: a new method for control of acute gastrointestinal bleeding.' *Radiology* 102: 303–306.
3. Nusbaum M, Baum S, Sakaiyalak P et al. (1967) 'Pharmacologic control of portal hypertension.' *Surgery* 62: 299–310.
4. Alavi A, Ring EJ (1981) 'Localization of gastrointestinal bleeding: superiority of ^{99m}Tc sulfur colloid compared with angiography.' *AJR* 137: 741–748.
5. Alavi A, Dann R, Staum M (1978) 'Efficiency of in vivo ^{99m}Tc red cell labeling.' Presented at The World Federation of Nuclear Medicine and Biology Meeting, Washington D.C., September 17–21.
6. Smith TD, Richards P (1976) 'A simple kit for the preparation of ^{99m}Tc-labeled red blood cells.' *J Nucl Med* 71: 126–132.
7. Winzelberg GG, McKusick KA, Froelich JW, Callahan RJ, Strauss HW (1982) 'Detection of gastrointestinal bleeding with ^{99m}Tc-labeled red blood cells.' *Semin Nucl Med* 12: 139–146.
8. Keller RT, Logan GM Jr (1976) 'Comparison of emergent endoscopy and upper gastrointestinal series radiography in acute upper gastrointestinal hemorrhage.' *Gut* 17: 180–184.
9. Sandlow LJ, Becker GH, Spelberg MA,, et al. (1974) 'A prospective randomized study of the management of upper gastrointestinal hemorrhage.' *Am J Gastroenterol* 61: 282–289.
10. Morris DW, Levine GM, Soloway RD, Miller WT, Marin GA (1975) 'Prospective randomized study of diagnosis and outcome in acute upper gastrointestinal bleeding: endoscopy versus conventional radiography.' *Am J Digest Dis* 20: 1102–1109.
11. Dronfeld MW, McIllmurray MB, Ferguson R, Atkinson M, Langman MJS (1977) 'A prospective, randomised study of endoscopy and radiology in acute upper-gastrointestinal bleeding.' *Lancet* 1: 1167–1169.
12. Graham DY (1981) 'Limited value of early endoscopy in the management of acute upper gastrointestinal bleeding: prospective controlled trial.' *N Engl J Med* 304: 925–929.
13. Alavi A: Unpublished data.
14. Alavi A, Dann RW, Baum S et al. (1977) 'Scintigraphic detection of acute gastrointestinal bleeding.' *Radiology* 124: 753–756.
15. Miskowiak J, Munck O, Nielson S et al. (1977) 'Abdominal scintigraphy with ^{99m}Tc-labeled albumin in acute gastrointestinal bleeding: an experimental study and a case-report.' *Lancet* 2: 852–854.
16. Winzelberg GG, Froelich JW, McKusick KA et al. (1981) 'Radionuclide localization of lower gastrointestinal hemorrhage.' *Radiology* 139: 465–469.
17. Markisz JA, Front D, Royal HD, Sacks B, Parker JA, Kolodny GM (1982) 'An evaluation of ^{99m}Tc-labeled red blood cell scintigraphy for the detection and localization of gastrointestinal bleeding sites.' *Gastroenterology* 83: 394–398.
18. Alavi A (1980) 'Scintigraphic demonstration of acute gastrointestinal bleeding.' *Radiology* 5: 205–208.
19. Alavi A, McLean GK (1980) 'Radioisotopic detection of gastrointestinal bleeding: an integrated approach to the diagnostic and therapeutic modalities.' In Freeman LM, Weissman H (eds), *Nuclear Medicine Annual 1980*. pp 177–218. New York: Raven Press.

20. Alavi, A (1982) 'Detection of gastrointestinal bleeding with ^{99m}Tc-sulfur colloid.' *Semin Nucl Med* 12: 126–138.
21. Alavi A. 'Scintigraphic detection and localization of gastrointestinal bleeding sites.' In Gottschalk A, Hoffer PB, Potchen EJ (eds) (1988) *Diagnostic Nuclear Medicine*. Baltimore: Williams & Wilkins, 1988 pp 631–662
22. Dann R, Alavi A, Baum R et al. (1980) 'A comparison of in vivo labeled red blood cells with Tc-sulfur colloid in the detection of acute gastrointestinal bleeding.' *J Nucl Med* 21: 75 (abstr).
23. Barth KH, Strandberg JD, White RI (1977) 'Long term follow-up of transcatheter embolization with autologous clot, Oxycel and Gelfoam in domestic swine.' *Invest Radiol* 12: 273–380.
24. White RI, Strandberg JD, Gross GS et al. (1977) 'Therapeutic emboliztion with long term occluding agents and their effects on embolized tissues.' *Radiology* 125: 677–687.
25. Gianturco C, Anderson JH, Wallace S (1975) 'Mechanical devices for arterial occlusion.' *AJR* 124: 428–435.
26. Anderson JH, Wallace S, Gianturco C et al. (1979) '"Mini" Gianturco stainless steel coils for transcatheter vascular occlusion.' *Radiology* 132: 301–303.
27. Youmans CR (1970) 'Cystoscopic control of gastric hemorrhage.' *Arch Surg* 100: 721–723.
28. Papp JP (1976) 'Endoscopic electrocoagulation of upper gastrointestinal hemorrhage.' *JAMA* 236: 2076–2079.

13. Intestinal absorption tests

RICHARD BERBERICH

Introduction

The absorption and excretion of radiopharmaceuticals is still of interest in diagnostic of gastrointestinal diseases with nuclear-medicine investigations. In this chapter the most common methods of measuring absorption and excretion are discussed. The performance of the different tests and their standard values are described. Moreover the basic possiblities of measuring absorption and excretion and the needed measurement equipments are presented.

The absorption and excretion tests in the nuclearmedicine diagnosis are of main interest in gastrointestinal and haematologic diseases. Table 1 shows an overview concerning the frequently performed nuclear-medicine absorption and excretion tests. Beside these described tests exists a lot of other tests for other radiopharmaceuticals labelled with β-emitters like fats, lipids, carbohydrates and cholic acids. For the measuring of the absorption and excretion of radiopharmaceuticals different methods and measurement equipments can be used. For this reason the principles of the different methods to perform absorption and excretion tests are described in the following. Afterwards the tests listed in Table 1 are described in detail with the range of here normal values.

Principles of absorption and excretion test 1

Excretion test with stool collections

With these tests one can estimate the absorption of a radiopharmaceutical by collecting the stool over a defined period and measuring its activity after administration of a oral test-activity. Performing these test is not very agreeable and in the case of radiopharmaceutical labelled with β-emitters often ash analysis is nesessary to measure the activity. For measuring the test and the stool activity suitable standards with comparable volumes must

H.J. Biersack and P.H. Cox (eds), Nuclear Medicine in Gasteroenterology, 191–199

Table 1. Absorption tests

Absorption tests in nuclear medicine gastroenterology		
Iron:	^{59}Fe	Chloride or sulphate
Vitaman B_{12}:	^{57}Co, ^{59}Co	Cyanocobolamin
Bile acids:	^{75}Se, ^{14}C	Homocholic-acid
		Conjugated with taurine
Measurement of enteral excretion		
Gastrointestinal protein loss		^{51}Cr-human serum albumin ^{51}Cr-^{12}C
Gastrointestinal blood loss		^{51}Cr-labelled erythrocytes
		^{111}In labelled erythrocytes

be available. In the case of radiopharmaceuticals labelled with Gamma-emitters the measurement of the activities can be done very easily with a well counter. In these tests it is of great importance to collect the stool over a sufficient long period. The faeces excretion tests need a good compliance between the staff of the hospital and the patient. Most of the difficulties in these investigations are due to the selective collection of the stool without any contamination of urine, especially if the radiopharmaceutical is labelled with an iodine-isotope.

Resorption tests by measuring blood and plasma activities

In these tests the time course of plasma and blood-activity is measured by a series of blood samples after oral application of the test activity. The steepness of the rise of the activity represents an indirect measure concerning the quickness of the absorption. But these tests allow only a qualitative judgement of the absorption. A quantitative evaluation is in most cases impossible since the distribution volumes and their exchange constants are not known. Another disadvantage of these tests is the fact that a lot of blood samples must be drawn often over a time period of several days.

Dual-isotope-methods

Performing the investigations with these procedures the radiopharmaceuticals must be labelled with two isotopes of different energies. One of the radiopharmaceuticals is then given orally and the other is injected intraveinously. After complete absorption an equilibrium between the two substances in whole blood is reached. The proportion between the resorbed activity and the intraveinously administered activity is then a measure of the absorbed part of the radiopharmaceutical. Because of the fact, that the time, when the equilibrium is reached, is unknown a lot of blood samples must be drawn. Therefore these tests are time consuming and unpleasant for the patients.

Measurements with a whole body counter

If radiopharmaceuticals labelled with Gamma-emitters are available for the absorption tests, they can be done with a whole body counter. After oral administration, the activity is measured two to three hours later under the assumption that at this time the radiopharmaceutical is distributed homogeneously in the body of the patient. Depending on the quickness of the absorption process of the used radiopharmaceuticals measurements must be carried out over several days or weeks.

The advantages of the absorption measurements with the whole-body-counter are the small quantities of activity needed for the investigations and a reduction of the radiation dose for the patient. Mistakes can occur if the radiopharmaceutical is not distributed homogeneously in the body of the patient. Another disadvantage are the high costs and the fact, that this equipment can only be used for few investigations. If a whole body counter is not available other measurement equipments like a two-probe-counter or an uncollimated Gamma-camera can be used instead. But there are some restrictions concerning their use. They have a lower sensitivity and it is often difficult to reach a constant sensivity over the trunk of the patient.

Exhalation measurements

If ^{14}C labelled organic substances are used the absorbed part of a radiopharmaceutical can be measured by exhalation. This is possible under the assumption, that after absorption Co_2 labelled with ^{14}C is produced by metabolic processes. This labelled Co_2 is then exhaled over the lungs. For this measurement procedure one believes, that the exhaled activity is related directly with the absorbed part of the radiopharmaceutical. The exhaled activity can be easily be measured with an ionisation chamber. The main advantage of the exhalation test is the fact, that stool collections can be avoided.

Absorption measurements by urinary output

Some radiopharmaceuticals are excreted after absorption by urinary output. The part of the substance excreted by urinary output is an indirect measure for its absorption. Difficulties in these tests may be due to the exact collection of urine. For these tests also a normal kidney function must be required.

Special methods

Gastro-intestinal-protein loss

The measurement of gastrointestinal protein loss is of great interest in lymphatic disorders, gastrointestinal tract ulcerations and in diseases with me-

Table 2. Detection of gastrointestinal protein loss

Excretion test for protein loss

1. Slowly intravenous injection of 9–11 mBq ^{51}Cr-chloride
2. Quantitative stool collections over a time period of 4 days (urinary contamination must be avoided absolutely)
3. Measurement of the stool activity and calculation of the excreted percentage of the given ^{51}CrCl activity

Normal values

Excretion less than 2% in a time period of 4 days.

Table 3. Measurement of iron absorption

Measurement of iron-absorption with the whole body counter

1. Oral application of 17.5–37 kBq ^{59}Fe-chloride
2. Measurement of the whole-body-activity 2 hours after administration
3. Measurements of the whole-body-activity after 7 and 14 days
4. Calculation of absorbed percentage

Values in percent of given activity

Normal males:	10–30%	Males with iron deficiency:	90–100%
Normal females:	20–40%	Females with iron deficiency:	70–100%
Female after cessation of the menses:	15–30%		

chanism of loss unknown. The procedure for this test is described in Table 2. Several other radiopharmaceuticals like ^{131}J or ^{125}J albumin [14, 15, 17]. ^{131}J Polyvinylpyrrolidone [3, 4] has been used. Due to urine contamination and excretion of free iodine by the salivary glands false positive results may occur. Now ^{51}Cr–Cl_2 or ^{51}Cr-albumin are widely used [14]. Another possibility is to perform the investigation with a ^{59}Fe-dextran complex [16]. In this case urine contamination cannot be expected. But the radiation dose for the patient is relatively high.

Iron-absorption-test

Table 3 shows the performance of the iron absorption measurement with the whole body counter [5, 18] together with the normal values for males and females. The iron absorption-test can also be done by stool collections over a period of 6 to 10 days. The long collection period is necessary since the passage through the intestine is very long. This is due to the fact that iron is bounded to the cells of the mucosa. Another disadvantage of this test is the high radiation dose due to the applied activity of 185 to 370 kBq ^{59}Fe-sulphate.

The measurement of iron absorption is of clinical interest in diseases like hypoferric anemia, haemochromatosis and haemosiderosis [6]. The absorption test should be combined with the determination of the total iron binding

Table 4. Absorption of Vitamin B_{12}

Dual-isotope-absorption test for Vitamin B_{12}

1. Oral application of 18.5 kBq ^{57}Co-cyanocobolamin bounded to intrinsic factor and 30 kBq ^{58}Co cyanocobolamin.
2. If the investigation is done with a whole-body-counter measurement of the whole-body-activity after 2 h and one week later for calculation of the absorbed part.
3. Schilling Test: Two hours after oral application intramuscular injection of a flushing dose of 1000 μg Vitamin B_{12}. Urine collection over a time period of 24 h with following measurement of the urine-activity and calculations of the excreted percentage of the activity.

Normal values of the B_{12}-absorption without intrinsic factor

1. Whole body counter: > 30% of the given activity
2. Schilling Test: > 10% Excretion of the given activity in urine collected for 24 h

Table 5. Measurement of bile acid absorption

Possibilities of measuring bile acid absorption

1. Oral application of 37–370 kBq ^{75}Se-homo-cholic acid conjugated with taurine under fasting condition
2. Measurement method for the absorption
   a) Determination of the relative uptake in the gall bladder after 6 h
   b) Determination of the biological half-life with the whole body counter or a comparable measurement equipment (p.e. an uncollimated gamma-camera)
   c) Measurmeent of the Retention value with a whole body counter 7 days after application.

Normal values

a) More than 80% of the activity in the gall bladder, if liver function is normal
b) Biological half-live greater 2.5 days
c) Absorption value bigger than 15% after 7 days.

capacity (TIBC), the unsaturated (latent) iron binding capacity (CIBC) and the plasma level of ferritin. If for a patient the determination of a ferrokinetic is planned, in cases of hypoferric anaemia and haemochromatosis the iron absorption test should be done first, to avoid an unnecessary radiation dose to the patient due to a ferrokinetic investigation with ^{59}Fe.

Vitamin B_{12} absorption test

Vitamin B_{12} deficiency can give rise to anaemia, thrombocytopenia, leukopenia and spinal cord degeneration. The causes of Vitamin B_{12} deficiency can be inadequate ingestion, or malabsorption (due to gastric abnormalities and due to intestinal malabsorption). Three possibilities exist for measuring Vitamin B_{12} absorption, the Schilling test by urine collection, the faeces excretion test and the determination with the whole-body counter. In Table 4 the Vitamin B_{12} absorption test with the whole body counter and the Schilling

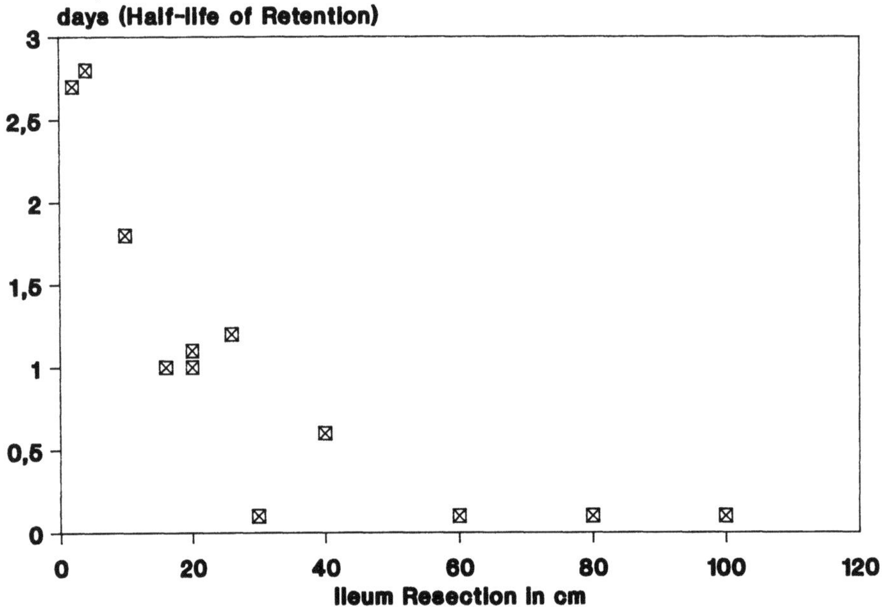

Figure 1. Correlation between the half life of retention of ^{75}Se-HCAT and the length of resection of the terminal ileum.

test [5, 7, 8, 11, 12] are described. The tests are described as dual-isotope-tests with ^{57}Co-cyanocobalamin [1]. Important for all tests is that the amount of administered stable Vitamin B_{12} must be smaller than 2 μg. If higher amounts are given the absorption is disturbed and false values are calculated. In cases of intrinsic factor deficiency the absorption is normal with ^{57}Co cyanocobalamin bounded to intrinsic factor. In intestinal malabsorption syndromes the absorption is reduced with both radiopharmaceuticals and in cases of inadequate ingestion both absorption values are normal.

Test for the absorption of homocholic acid

In Table 5 the possiblities for the determination of the absorption of homocholic acid are shown. Due to the development of Se-75 homocholic acid conjugated with taurine (Se-75-HCAT) the absorption measurement of cholic acid has become easier to perform. Formerly only cholic acids labelled with β-emitters could be used. The absorption test of cholic acid is of great interest in the differential diagnosis of chronic diarrhoea. One of the causes of chronic diarrhoea is the enteric loss of bile acids, which can occur as a result of inflammatory infection of the terminal ileum. Crohn's syndrome a chronic inflammatory disorder of the intestinal tract is found all over the world with different incidence of 2 to 40 per 100 000 [9].

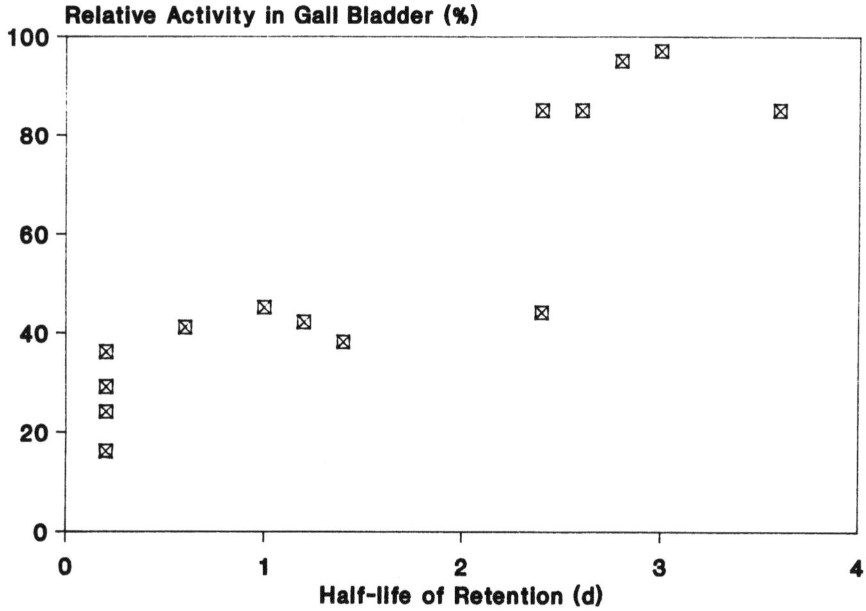

Figure 2. Correlation between the gall bladder activity after 6 hours and the half life of retention measured with the whole body counter of ^{75}Se-HCAT.

For the measurement of the bile acid absorption a whole body counter or a similar equipment like an uncollimated Gamma-camera can be used [2, 10, 13]. In case of the whole body counter 37 kBq and in the case of uncollimated Gamma-camera 370 kBq and ^{57}Se-HCAT are administered orally under fasting condition. The activity is then registrated after 2, 4 and 8 hours. Corresponding to the decrease of activity measurement are continued over a time period of 4 days once in the morning and once in the evening. During the investigation the patient must eat normal food. After the end of the procedure the activity values are approximated by monoexponential fit. In normal subjects the biological half-life of the bile acid is greater than 2.5 days. In patients with a Crohn's syndrome of the terminal ileum the half-life is often shortened to values lower than 0.5 days. In patients where a part of the terminal ileum was resected, the length of the removed part corresponded well with the reduction of the bile acid absorption. This correlation is shown in Fig. 1. If the resection length was greater than 40 cm the half life of the ^{75}Se-HCAT was always shorter than 0.5 days. The determination of bile acid absorption can also be done with the whole body counter with two measurements after distribution in the body and one week later. Values higher as 15% proof an undisturbed absorption. The third possibility to determine bile acid absorption is the measurement of the relative gall-bladder-activity with a Gamma-camera six hours after oral application. Values over a level of 80% proof an undisturbed absorption. The procedure is the following: The patient receives 370 kBq ^{75}Se-HCAT orally. If the capsule has reached the stomach scintigrams in ventral and dorsal view are taken from this region. With regions of interest the geometrical mean of the count-rates of the capsule are determined. This measurement is necessary to adjust the measurement of the gall bladder activity under the assumption that the measurement geometry is the same. Six hours after the administration of the capsule the activity of the gall bladder is measured in ventral and dorsal view and then compared with the given activity. Figure 2 shows the comparison between the absorption values measured with the whole body counter and the relative activity in the gall bladder after 6 h. There is a good correlation between both measurement procedures. The advantage of measuring the relative gall bladder activity is the quick performance of the absorption test, but the measurement is only possible in patients with normal liver and gall bladder function. Though if the relative gall bladder activity is greater then 80 percent after 6 hours an absorption disturbtion can be excluded. If the value is smaller, the measurement must be continued with the whole body counter.

References

1. Bayly RJ, Bell TK, Waters A (1971) 'A dual isotope modification of the Schilling-test,' in: *Ergebnisse der klinischen Nuklearmedizin*. Schattauer-Verlag: 911–915.
2. Boyd GS, Merrick MV, Monks R, Thomas I (1981) '^{75}Se-labelled bile acids, new radiopharmaceuticals for investigating the enterohepathic circulation.' *J Nucl Med* 22: 720–725.
3. French AB, et al. (1961) 'Distribution of ^{131}J polyvinylpyrrolidone in human after oral and intravenous administration.' *Fed Proc* 20: 242.
4. Gordon RS Jr (1959) 'Exudative enteropathy: abnormal permeability of the gastrointestinal tract demonstrable with labelled polyvinylpyrrolidone.' *Lancet* I: 325–326.
5. Heinrich H (1967) 'Gesamtkörper-Radioaktivitätsmessung.' *Therapiewoche* 51: 113.
6. Heinrich H, Bartels H (1967) 'Bestimmung, Methoden und Normalbereiche der intestinalen Eisenresorption beim Menschen.' *Klin Wschr* 45: 553.
7. Heinrich HC (1961) 'Radio-Vitamin B_{12} in der klinischen Diagnostik. Künstliche radioaktive Isotope in Physiologie, Diagnostik und Thereapie. (Hrsg Schwiegk H, Turba F.). 2. Aufl Bd II S 660. Berlin–Heidelberg–New York: Springer.
8. Kunkel R, Schatanek W, Oberhausen E (1972) 'Ergebnis der Resorptionsmessungen von ^{58}Co-Vitamin B_{12} mit dem Ganzkörperzähler.' Nuklearmedizin, klinische Leistungsfähigkeit und techn. Entwicklung. S 301–303. Stuttgart: KF Schattauer.
9. Lorenz-Meyer H, Brandes W (1981) 'Klinik des M. Crohn.' *Internist* (*Berlin*) 22: 420–429.
10. Nyhlin H, Merrick MV, Eastwood MA, Brydon WG (1983) 'Evaluation of illegal function using 23-selena-25 homotaurocholate, a gamma labelled conjugated bile acid.' *Gastroenterologie* 84: 63–68.
11. Pribilla W (1967) 'Der Schilling-Test.' *Dtsch med Wschr* 92: 1774–1776.
12. Schilling RF (1953) 'Intrinsic factor studies II. The effect of gastric juice on the urinary excretion of radioactivity after the oral administration of radioactive vitimin B_{12}.' *J Lab Clin Medicine* 42: 860–866.

13. Schroth HJ, Berberich R, Feifel G, Müller KP, Ecker KW (1985) 'Tests for the absorption of (^{75}Se-HCAT) Tauro-23-Selenium-25 Homocholic acid.' *Eur J Nucl Med* 10: 455–457.
14. Waldmann TA (1961) 'Gastrointestinal protein loss demonstrated by ^{51}Cr-labelled albumin.' *Lancet* 2: 121.
15. Waldman TA (1966) 'Protein losing enteropathy.' *Gastroenterology* 50: 422.
16. Waldmann TA (1969) 'Protein-losing enteropathy.' *Mod Trends Gasteroenterol* 4: 125.
17. Wochner RD et al. (1968) 'Direct measurement of the rates of synthesis of plasma proteins in control subjects and patients with gastrointestinal protein loss.' *J Clin Invest* 47: 3.
18. Wolf F (1979) 'Gastroenterologie, Emrich, D., Nuklearmedizin Funktionsdiagnostik und Therapie.' Georg Thieme Verlag, Stuttgart, S. 280–317.

PART THREE

Miscellaneous

14. Investigations of disorders of motility of the esophagus in chronic diseases

W. MECKLENBECK and HENNING VOSBERG

Abstract

Chronic diseases of the esophagus impair the transport function of this organ. Esophageal scintigraphy investigates the transport function under physiological circumstances. Various test meals are labelled radioactively and swallowed by the patient. Depending on the cause of a transport disorder the investigation lasts only 60 s up to half an hour. Parametric imaging techniques like the so called condensed pictures give informations about extent and peristalsis of transport disorders. The determination of transit times and/or percent of residuum in the esophagus allows to quantitate functional disorders. The use of the multiple swallow technique or of various tracers during one investigation leads to a high sensitivity in the detection of esophageal transport disorders, even in an early stage of a chronic disease, like morphea or scleroderma. The documentation of successful therapeutical interventions is possible, for example in achalasia. In obscure complaints or in thoracic pain without coronary heart disease esophageal scintigraphy may confirm or exclude a disease of the esophagus.

Introduction

Various diseases of the esophagus show similar clinical signs such as dysphagia, sometimes with regurgitation, retrosternal pressure and pain. As all diseases of the esophagus impair the transport function of this organ in some way, assessment of this function is of clinical importance. Quantitative measurements should be carried out so that the results of repeated investigations in the course of time may be compared. This is important because of the primarily chronic character of many deseases of the esophagus. In addition to the documentation of pharmacological, endoscopic and surgical therapy effects, the special advantage of investigations by methods of nuclear medicine is that they are noninvasive and can be performed under physiologi-

H.J. Biersack and P.H. Cox (eds), Nuclear Medicine in Gasteroenterology, 203–216

cal conditions without irritation by endoscopic instruments or aversion to test substances.

Methods

Composition of radioactively labelled tracers

In order to investigate esophageal transport function the patient swallows suitable food labelled with radioactive tracers. Immediately afterwards monitoring with a gamma camera-computer system begins. The results of the investigation depend first of all on the kind and quantity of the food. Normally, the investigation starts with a liquid tracer such as water or juice. The patient is asked to swallow a spoonful of liquid completely and then to swallow several times in the absence of liquid-dry [1, 2]. In normal subjects the transport time is 5–10 seconds from mouth into stomach. Several authors found a number of false normal results when using only a liquid tracer [3]. To improve the sensitivity of this investigation the patient may swallow several spoonfuls of a liquid tracer one after another so that one can observe whether a deterioration or an improvement of the transport function results [4, 5]. Alternatively it is possible to administer a semiliquid or solid tracer for a second investigation immediately afterwards. As a semiliquid tracer, mashed potatoes or curd, weighing about ten grams, may be used. The transport of semiliquid food requires more time than that of a liquid tracer [6, 7]. Moreover, semiliquid food is subdivided into several portions after swallowing and will be transported in several steps into the stomach [8]. One may use radioactively labelled parts of a pancake or bread in the form of small cubes of about 1 cm as a solid tracer.

Peristalsis is observed much better with solid tracers than with liquid or semiliquid tracers. After swallowing solid tracers only a small amount of percent residuum of the bolus (pr) remains in the esophagus. Liquid and semiliquid food can stick to the folds of the mucosal membrane of the esophagus and will clear only slowly [9]. Some particular questions can be answered by investigation with radioactively labelled gelatine capsules, simulating swallowed drugs [10].

To obtain complete information regarding the transport function of the esophagus, it is advisable to utilize several tracers of different consistency, thus more closely approximating physiological conditions.

Labelling materials

For radioactive labelling the radioactive tracer is simply mixed with the liquids and semiliquids; for labelling of bread or pancake the radioactive substance should be mixed with the dough. The radioactivity must be distributed homogeneously in the test meal. On this condition one may consider the amount of radioactivity in the esophagus to be equal to its volume.

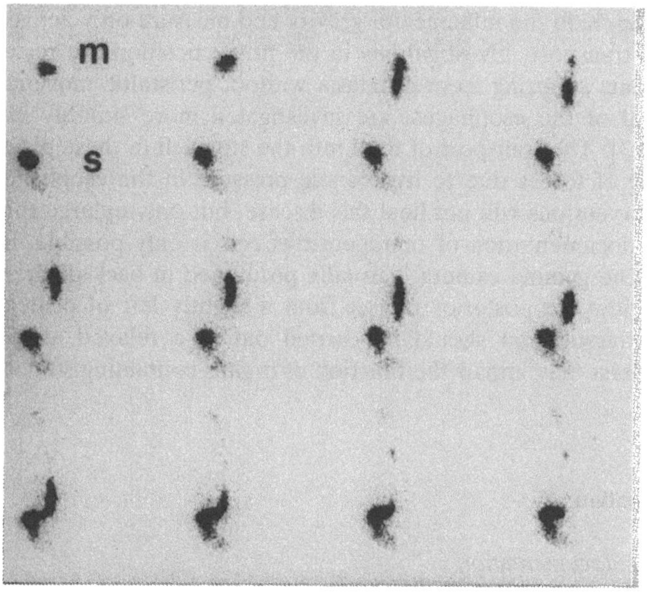

Figure 1. Serial scintiphotos of an esophageal scintigraphy with mashed potatoes; the sequence lasted 36 sec. The radioactivity is in the first picture in mouth (m) and stomach (s) and passes through the esophagus uninhibited.

For labelling ^{99m}Tc sulfocolloid or ^{99m}Tc-DTPA is suitable. Labelling with sulfocolloid has the advantage that it is not resorbed by the gastrointestinal tract; when using DTPA, the activity resorbed in the gastrointestinal tract will be excreted rapidly by the kidneys. The radiation dose to which the patient is exposed in either case is very low, as only 10–30 MBQ ^{99m}Tc-DTPA or sulfocolloid are used in each investigation.

Registration of data

Immediately after the test meal, data acquisition is started. With a gamma camera-computer system a rapid sequence of frames is obtained. For normal function of the esophagus, an acquisition of 60 frames per minute is sufficient. In the case of transport disorders of moderate severity the collection of data should be continued for 2 or 3 min. In transport disorders of greater severity longer acquisition times of up to half an hour are necessary. In these special cases frame sequences of 4 frames per minute are sufficient.

The transport function of the esophagus can be investigated with the patient in either supine or prone position. Some authors favour the investigation in a supine position as this is similar to physiological conditions [11]. But swallowing in the supine position represents a mixture of active peristaltic transport of food and a more passive mechanism, depending on gravitational force. To exclude the influence of gravity and measure only active muscular peristaltic transport, investigations in the prone position are recommended [12]. Patients suffering from achalasia without peristaltic movement in the lower third of the esophagus, are investigated more suitably in a supine position [13]. The transport of food into the stomach in these patients is the result only of forces due to hydrostatic pressure in the esophagus. Therapeutic interventions will not heal this disease, but only enlarge the stenosis, therefore documentation of therapeutic effects is only possible in a supine position. The gamma camera is usually positioned in back of the patient in order to view the posterior thorax from a slightly left of center position. Such an investigation should be carried out in a relaxed atmosphere as external stress may impair the function of organs containing smooth muscles [14].

Data evaluation

Parametric documentation

Sequential scintiphotos taken every one or two seconds following the swallow of a tracer offer a widely used and simple kind of documentation. Unfortunately documentation seems to be impossible in the presence of severe disorders of transport and even an accumulation of about 60 or 120 scintiphotos would not suffice. In addition such a procedure would be expensive. Monitoring by parametric imaging or by time-activity curves obtained by the ROI technique is more convenient. For this purpose an accepted procedure for parametric imaging is the acquisition of the so-called condensed pictures [15, 16]. Every single frame of a sequence study is condensed into a vertical stripe with a breadth of one pixel. Each stripe shows a one-dimensional distribution of the radioactivity between pharynx and stomach. Corresponding to the single frames of the sequence, the single stripes are connected side by side. Condensed pictures, therefore, relate time on the x-axis and of transport of activity on the y-axis.

Additionally, time-activity curves for the entire esophagus or for the upper, middly and lower third of this organ can be derived by means of the ROI-technique. The type and severity of a functional disorder may be derived from the shape of the curves [17]. There are several other parametric methods of studying the transport function of the esophagus, but because of their complex nature these methods are still not in common use [18, 19, 20].

Furthermore most of these methods give less information about peristaltic movement.

Quantitative results

It is, in principle, possible to calculate a minimal or mean transport time or a percent of residuum of the bolus (pr) in the esophagus [21].

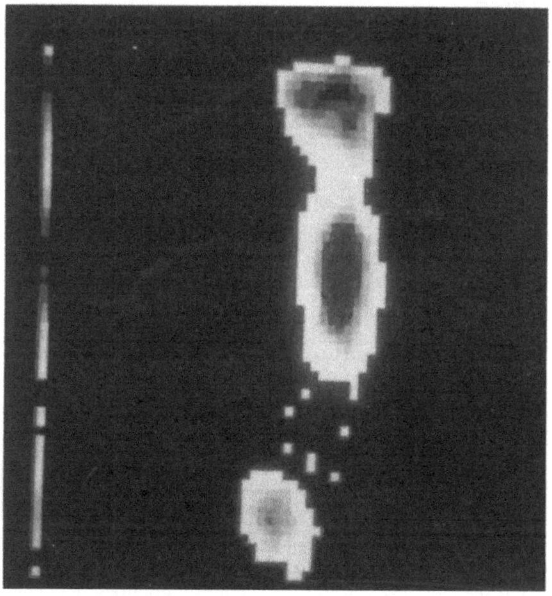

Figure 2. The right side of the figure shows a usual single frame out of an esophageal scintigraphic study. On the left side the frame is condensed to one stripe. Many of these stripes build up a condensed picture.

Minimal transport time is defined as the time difference between the beginning of the swallow and the appearance of the first radioactivity in the stomach. The mean transport time may be described as the time interval elapsing between the beginning of the swallow and the time of a 50% clearance of the material in the esophagus. Since the radioactivity of the entire bolus is set as 100%, the assessment of the percentage of residual of the bolus (pr) can be done at any point during the investigation.

In our investigations pr usually is determined after 30, 60, 120 and 240 s. In severe transport disorders it may be necessary to determine pr after 10, 20 and 30 minutes too. This method is comparable to the calculation of the percent esophageal transit, proposed by other authors [2].

Results

Percent residuums of radioactivity

The transport function of the esophagus was investigated in patients with so-called collagen diseases (scleroderma, morphea and certain forms of lupus erythematosus), with functional disorders (diffuse esophageal spasm, achalasia, stricture of the esophagus and gastroesophageal reflux). Normally swallowing of the various tracers occurs very rapidly. Table 1 shows normal

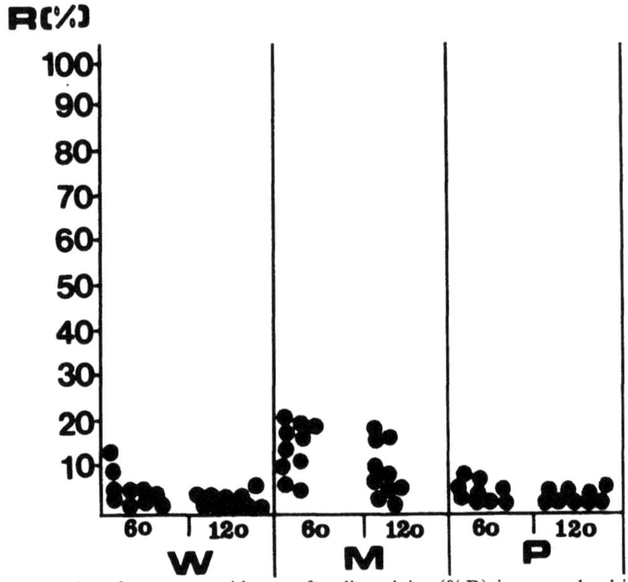

Table 1. Shows the results of percent residuum of radioactivity (%R) in normal subjects after 60 and 120 seconds. (W = water; M = mashed potatoes; P = pancake)

values for liquid, semiliquid and solid tracers. The normally remaining pr especially that from liquid and semiliquid tracers does not indicate a minimal functional disorder, even though the pr remains in the esophagus for a few minutes without further transport. This small fraction is probably without meaning for the transport function and can be explained for example by adherence of activity to the mucous membrane or in folds. A second swallow of unlabelled water clears the esophagus from these small residuums.

Collagen diseases

Among 59 patients with morphea and scleroderma we found 35 patients with increased values for pr after the liquid swallow. Among these patients were 23 without subjective complaints regarding swallowing. The percentage of food remaining in the esophagus in the case of a semiliquid tracer (mashed potatoes) was pathologically increased in 48 among 59 patients, (36 without subjective complaints regarding swallowing). Regarding swallowing complaints all patients but one showed an increased pr for water and mashed potatoes. This patient with morphea complained about swallowing disorders in the upper third of the esophagus, but no functional disorder could be established. Five patients with lupus erythematosus exhibited normal values. In five patients we found for pr values of mashed potatoes increased and in four patients also for water.

Achalasia and other diseases

We found the largest pr in patients suffering from advanced achalasia, who even after 30 min retained up to 90% of the swallowed liquid in the eso-

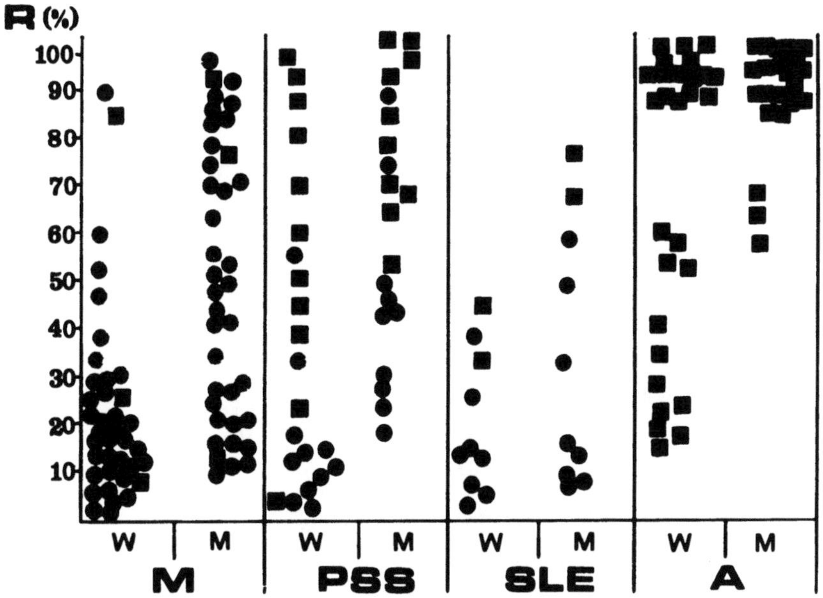

Table 2. Shows the results of pr (%R) in several groups of patients after 60 sec for water, the round points representing patients without the square points patients with swallowing complaints. (M = mophea; PSS = scleroderma; SLE = lupus erythematosus; A = achalasia)

phagus. Among 154 patients investigated 31 presented with a spontaneous gastroesophageal reflux some only once some repeatedly during the investigation. Only nine of these patients suffered from refluxesophagitis. Aside from reflux there was a significant increase in pr in 6 of these patients.

Among 21 patients with obscure chest pain four had severe and six mild transport disorders. One patient had a scar stricture of the esophagus as the cause of a severe transport disorder, whereas three others suffered from diffuse esophageal spasm.

The findings regarding the pr in the esophagus after 60 s are dispersed widely in the different patient groups (for example in collagen diseases and achalasia: see Table 2).

Transit times

The minimal and mean transit times were determined for all patients. Patients with collagen diseases showed similar mean transit times to the results gained by the documentation of pr. In severe transport disorders such as in achalasia or in scleroderma the determination of a mean transit time is possible only with difficulty since it requires a protracted time of measurement. Even in normal persons the determination is problematical as the activity curve does

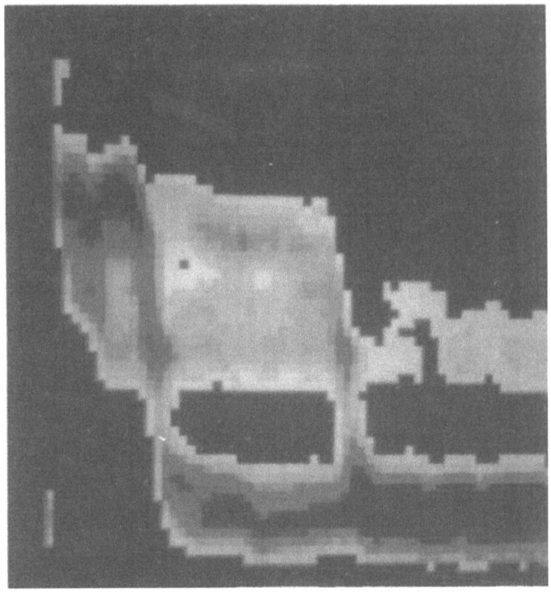

Figure 3. The condensed picture shows typical transport disorders in scleroderma. Transport in the upper and the lower third of the esophagus seems to be normal, in the middle third transport is delayed and the bolus is subdivided into several parts.

not return to the baseline, because of some physiological pr. In order to solve the problem of non-return to baseline we used a suitable mathematical extrapolation [22, 23]. Evaluation of the minimal transit time yielded disappointing results, because the transit time often turned out to be within normal limits, even in patients with severe transport disorders. In such patients small amounts of activity passed uninhibited through the esophagus, while a larger amount remained in the esophagus.

Condensed pictures

Statements regarding the nature and location of a transport disorder are possible by means of the condensed picture method mentioned above. In healthy persons esophageal transport of water or mashed potatoes is accomplished by few peristaltic waves. The transport of mashed potatoes is always slower than that of water. With a suitable spatial resolution, faster and slower components of transport in the middle and the lower third can be distinguished [16].

However solid tracers follow an all-or-none rule since they do not disintegrate and are transported as a compact mass through the esophagus without a large residuum.

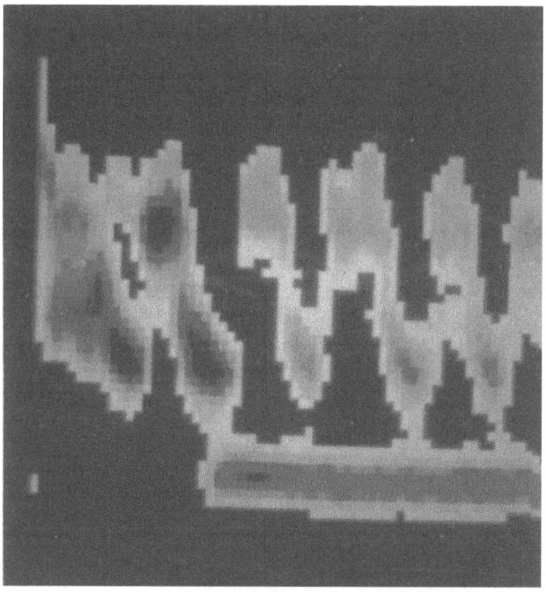

Figure 4. The condensed picture shows transport disorders in the middle third of the esophagus. There is a remarkable peristalsis including nearly the whole esophagus. The Figure shows a typical prestenotic peristalsis in a patient suffering from a cardiatumor.

Collagen diseases

Collagen diseases such as morphea and scleroderma usually lead to transport disorders in the middle and lower third of the esophagus, but seldom in the upper third [24]. In these patients transport is accomplished in several steps associated with a prolonged transport time. In scleroderma transport disorders are often severe in comparison to other collagen diseases.

Achalasia

In achalasia there is often minimal transport in the lower third of the esophagus; only two patients in our series showed transport disorders in the middle third. The initial transport from pharynx to the middle of the esophagus proceeds normally. Achalasia characteristically lacks an active prestenotic peristalsis, since the primary lesion is an innervative deficit [25]. In 31 patients with achalasia we found only one with an active peristalsis.

Other diseases

Mechanical stenosis due to scar tissue, strictures and tumors can be found in every part of the esophagus and usually show a strong prestenotic peristalsis. In all groups of patients and even in normal subjects we found some with spontaneous gastroesophageal reflux, which can be visualized on the

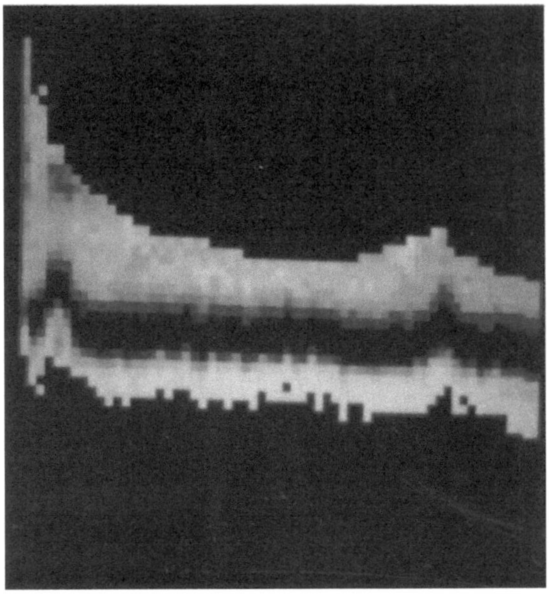

Figure 5. The condensed picture shows a normal transport in the upper third of the esophagus and nearly no transport in the middle third without vivid peristalsis in a patient suffering from achalasia.

condensed pictures as a rise in activity in the lower and middle third of the esophagus ascending from the stomach. In 6 patients with refluxesophagitis we found mild transport disorders in the lower third of the esophagus. A significant difference between spontaneous reflux in normal subjects and a reflux leading to esophagitis could not be ascertained [26, 27, 28]. In this entire group of patients, we found transport disorders in all parts of the esophagus depending on the cause of the disease.

Discussion

Condensed pictures

The documentation of the esophageal function by condensed pictures is a substitute for the infinitely more difficult presentation of numerous single scintiphotos. The transport of activity through the esophagus during the course of time is presented easily, understandably, and clearly. There exist typical patterns for hypermotility (stenosis), hypomotility (collagen disease) and amotility (achalasia) transport disorders. Localisation and extent of a functional disorder, artifacts like a second swallow or a reflux can be well determined. A more precise differentiation as between organic stenosis and the presence of scars or tumors is not possible.

Time-activity curves

Time-activity curves encompassing over the whole or several parts of the esophagus are the basic to every quantitative determination. If intra- and interindividual comparison are desired, it is useful to quantify esophageal function. The diagnosis of the kind and precise localisation of a disorder by time-activity curves is more difficult [29, 30]. Although there are typical changes of such curves, e.g., a plateau in stenosis, peristaltic dynamics cannot be easily elucidated.

Transport times

The estimation of transport times is more difficult. Minimal transport time from pharynx to stomach may be normal even in severe transport disorders since small parts of the bolus pass the esophagus rapidly without delay. The determination of a mean transit time can be difficult in severe transport disorders such as achalasia, when there is still more than 50% of the radioactivity remaining in the esophagus after 30 min. Moreover, the mean transit time constitutes an average value for transport and physiological pr. The latter may distort the results referring to the relatively broad distribution of the normal values of physiological pr. Since the esophageal time activity-curves do not show a one exponentional slope, the exact determination of a mean transit time may rise problems. The measurement of the pr is very useful in intra- and interindividual comparisons and even results obtained by different authors can be compared if the bolus is of the same composition. Problems may arise if solid tracers like bread and pancake are used. Either the nearly whole bolus remains in the esophagus or it has been transported into the stomach nearly completely. When using solid tracers the determination of a mean transit time seems to be more appropriate than that of the pr. Gastroesophageal reflux or a second swallow of residuals from the pharynx cause interference with the determinations of pr or a mean transit time. These problems can be resolved by means of the condensed pictures. Although a spontaneous reflux is detectible on condensed pictures clinical conclusions are more difficult. We found transport disorders in most patients with reflux esophagitis, but there is no typical picture establishing the existence of an inflammation at the lower esophagus due to reflux [28].

The measurement of the transport function of the esophagus by means of radioactively labelled liquid, semiliquid and solid tracers represents a physiological and simple method for the detection of disorders of the esophagus. Above all, such measurements permit an early diagnosis in chronic diseases. When using various tracers or the method of multiple swallows false negative results are less often seen than in esophageal manometry, ph measurements and endoscopy [24, 28, 31].

A problem in esophagus scintigraphy derives from possibly false positive results, showing only mild pathological values without subjective complaints or other pathological results in other investigations [25]. For example we found among patients with and without coronary heart disease but with thoracic pain several moderate but clearly pathological values which remained unchanged with time. No other investigative method could detect these small transport disorders, as it is not possible to prove functional disorders even by macro- or microanatomical investigations. Up to this time, transport disorders of the esophagus in morphea were thought to be nonexistent, although several authors found results similar to ours [32, 33]. Only the further progression of these diseases will show the relevance of such investigations.

The examination of patients with collagen diseases showed that it is possible to detect diseases of the esophagus at a very early stage and to document the progress and the severity by functional scintigraphy [13, 33]. Esophageal scintigraphy is a suitable screening method for confirmance or excluding transport disorder in obscure swallowing complaints, neurological diseases and thoracic pain without coronary heart disease [35, 37]. In patients with achalasia esophageal scintigraphy provides evidence of therapeutic improvements [38, 40].

Although functional disorders of the esophagus can be detected by means of esophageal scintigraphy, the cause of these disturbances must be further investigated by means of radiological methods and endoscopy.

References

1. Büll U, Hör G (1987) *Klinische Nuklearmedizin* Edition Medizin 170.
2. Harbert J, Da Rocha A (1984) *Textbook of Nuclear Medicine* Vol 2 Lea and Febiger 217–220.
3. Holloway R, Lange R. Magyar R, Greene, McCallum R (1985) 'Radionuclide Esophageal Transit of a Liquid Bolus' *J Nucl Med* 25: 12.
4. Klein H, Wald A (1987) 'Normal Variation in Radionuclide Esophageal Transit Studies' *Eur J Nucl Med* 13: 115–120.
5. Ham H, Georges B, Froideville J, Piepsz A (19??) 'Esophageal Transit of Liquid: Effects of Single or Multiple Swallows' *Nucl Med Comm* 6: 263–267.
6. Blackwell J, Hannan W, Adam R, Heading C (19??) 'Radionuclide Transit Studies in the Detection of Esophageal Dysmotility' *Gut* 24: 421–426.
7. Mecklenbeck W, Frick-Gipp C, Vosberg H, Szabo Z (1987) 'Die Ösophagusfunktionsszintigraphie zur Diagnostik und Verlaufsbeobachtung Chronischer Erkrankungen mit Ösophagusbeteiligung' *Nuclearmedizin* 2a: 39–40.
8. Mecklenbeck W, Vosberg H (1988) 'Die Ösophagusfundtionsszintigraphie bei Chronischen Inneren Erkrankungen mit Ösophagusbeteiligung' *Der Nuklearmediziner* 5: 339–345.
9. Fisher R, Mahmud L, Applegate G, Rock E, Lorger St (1982) 'Effect of Bolus Composition on Esophageal Transit: Concise Communication' *J Nucl Med* 23: 878–882.
10. Malmud L, Rock E, Applegate G, Reilly J, Fisher R (1981) 'Esophageal Transit of Gelatin Capsules – Therapeutic Implications' *J Nucl Med* 22: 66.

11. Krosin G, Saladino T, McCallum R (1981) 'Radionuclide Quantification of Esophageal Emptying of Solid Food in Achalasia' *J Nucl Med* 22: 66.
12. Ryan J, Brunsden B, O'Sullivan G, Kirchner P, De Meester S, Cooper M (1981) 'The Measurement of Esophageal Motor Function' *J Nucl Med* 22: 28.
13. Drane W, Karvelis K, Johnson D, Curran J, Silverman E (1987) 'Szintigraphic Detection of Metoclopramide Esophageal Stimulation in Progressive Syst Sclerosis' *J Nucl Med* 28: 810–815.
14. Anderson K, Dalton C, Bradley L, Richter J (1989) 'Stress Induces Alteration of Esophageal Pressures in Healthy Volunteers and Non Cardiac Chest Pain Patients' *Dig Dis a Science* 34: 83–91.
15. Svedberg J (19??) 'The Bolus Transport Diagram: A Functional Display Method Applied to Esophagus Studies' *Clin PhysPhysiol Meas* 3: 267–272.
16. Klein H (1984) 'Esopohageal Motility: Computer Analysis of Radionuclide Esophageal Transit Studies' *J Nucl Med* 25: 957–969.
17. Espinola D, Yang P, Engin S, Douglas K, Loudenslager D, Wagner H (1983) 'A Characteristic Pattern of Esophageal Emptying in Obstructive Lesions of the Esophagus' *J Nucl Med* 24: 39.
18. Tatsch K, Knesewitsch P, Leisner B (1987) 'Visualisation of Esophageal Motility and Disorders by Parametric Imaging: A New Diagnostic Approach' *Nuklearmedizin* 4: 79.
19. Ham H, Georges B, Verelst J, Cadranel S, Piepsy A (1984) 'Assessment of Esophageal Transit in Adults and Children by Means $^{81\,m\,Kr}$' *J Nucl Med* 24: 98.
20. Tatsch K, Knesewitsch P, Kirsch C, Moser E (1988) 'Ösophagusszintigraphie Mit Parametrischen Bildern: Ein Neues Verfahren für die Diagnostik von Motilitätsstörungen' *Nuklearmedizin* 2: 17.
21. Ham H, Georges B, Froideville J, Piepsz A (19??) 'Esophageal Transit of Liquid: Effect of Single or Multiple Swallows' *Nucl Med Commun* 6: 263–267.
22. Ham H, Piepsz A (1985) 'Computer Analysis of Radionuclide Esophageal Transit Studies' *J Nucl Med* 26: 429.
23. Hanm H, Georges B, Guillame M et al (1984) 'Assessment of Esophageal Transit: Comparison of Several Data Processing Techniques' *Eur J Nucl Med* 9: 75.
24. Drane W, Karveli K, Johnson et al (1986) 'Radionuclide Esophageal Szintigraphy and Manometry in PSS' *Rad* 160: 73–76.
25. Datz F (1986) 'The Role of Radionuclide Studies in Esophageal Disease' *J Nucl Med* 25: 1040–1045.
26. Winzelberg G, Malik R, Ismail-Beigi F, Ennis M, Bruno S, Hershenson L (1983) 'Correlative Study of Liquid Phase Radionuclide Esophageal Transit Scintigraphy with Radiographic, Manometric and Endoscopic Parameters in Patients with Suspected Esophagitis' *J Nucl Med* 24: 39.
27. Singh A, Wang G, Holmes R, Butt J, Welch S (1985) 'Esophageal Emptying in Esophageal Disorders' *J Nucl Med* 26: 855–862.
28. Heyman S (1987) 'Gastric Emptying, Gastroesophageal Reflux and Esophageal Motility. Effective Use of Computers in Nuclear Medicine' *Gelfand a Thomas* 413–437.
29. Tolin R, Malmud L, Reilly R, Fischer S (19??) 'Esophageal Scintigraphy to Quantitate Esophageal Transit' *Gastroenterology* 76: 1402–1408.
30. Russel C, Holmes E, Hill G et al (1981) 'Radionuclide Transit: A Sensitive Screening Test for Esophageal Dysfunction' *Gastroenterology* 80: 887.
31. Russell C, Holmes E, Allen F, Hull D, Hill L (1987) 'Radionuclide Transit—Non Invasive Test for Esophageal Motility Disorder' *J Nucl Med* 22: 27.
32. Luderschmidt C, Leisner B, König G (1985) 'Die Ösophagusfunktionsszintigraphie als Parameter für die Interne Manifestation bei PSS' *Dermatologie und Nuklearmedizin* Springer Verlag.
33. Leisner B, Hundegger K, Luderschmidt C, König G (1985) 'Ösophagusfunktionsszintigraphie bei Kollagenosen. Dermatologie und Nuklearmedizin' Springer Verlag.
34. Tatsch K, Schröttle C, Kirsch C (1989) 'A New Combined Quantitative and Qualitative Parametric Multiple Swallow Test for Advanced Imaging of Esophageal Motility Disorders' *Nuclear Med* 15: 125.

35. O'Donnell J, Williams E, Nielson P, Hancock J, Benjamin S, Castell D (1982) 'Transit in the Evolution of Non Cardiac Chest Pain' *J Nucl Med* 23: 51.
36. Sand A, Lieber S, Roland J, Francken P, Vandevivere J (1987) 'Esophageal Transit in Atypical Thoracal Pain by Means of ^{81m}Kr *Nuklearmedizin* 4: 228.
37. Brand D, Martin D, Pope C (1977) 'Esophageal Manometrics in Patients with Angina-Like Chest Pain' *Am J Dig Dis* 23: 300–304.
38. Espinola D, Yang P, Loudenslager D, Koller D, Reepain H, Camargo E, Hendrix T, Wagner H (19??) 'Stimulation of Esophageal Emptying in Patients with Achalasia' *J Nucl Med* 23: 51.
39. Holloway R, Krosin G, Lange R et al (1983) 'Radionuclide Esophageal Emptying of a Solid Meal to Quantitate Results of Therapy in Achalasia' *Gastr* 84: 771–776.
40. Leisner B, Antes W, Brückner L (1978) 'Funktionsszintigraphie des Ösophagus am Beispiel der Operierten Achalasie' *Ergebnisse der Gastroent* 76.

15. Radioimmunoscintigraphy in gastroenterology

MARIA GRANOWSKA and KEITH BRITTON

Abstract

The development, general principles and clinical application of radioimmunoscintigraphy to primary and recurrent colorectal cancer is described. The types of monoclonal antibody, anti CEA, B72.3 or PR1A3 are compared and the importance of a good signal through the choice of the radiolabel is explained. The results from ^{111}In labelled and ^{99m}Tc labelled antibodies are presented in a clinical context.

Principles of radioimmunoscintigraphy

Radioimmunoscintigraphy is a technique developed through the cooperation of many disciplines whereby a monoclonal antibody is produced against an antigen on a target cell of usually a tumour, that antibody is radiolabelled so as to preserve its immunoreactivity and produced meeting radiopharmaceutical quality assurance standards, injected intravenously into a patient and its distribution imaged serially using a gamma camera so as to demonstrate sites of specific uptake.

Radioimmunotherapy uses the same approach for treatment. Nuclear medicine techniques generally have high sensitivity but low specificity. The only real advantage that a radiolabelled antibody has over the conventional radiopharmaceutical of nuclear medicine is the high specificity of the antigen-antibody interaction. However, to use that high specificity in order to undertake 'tissue characterisation' requires some preconditions to be met. A monoclonal antibody is a protein and its active fragments Fab and F(ab')2 are large peptides. Tumours have permeable capillaries so that any protein or peptide may diffuse in nonspecifically. The larger the tumour the lower the proportions of tumour to non tumour stroma and the greater the tendency to central necrosis. All these factors reduce the advantages of an antibody's specificity to that of a non specific tumour imaging agent. It is therefore evident that the technique of radioimmunoscintigraphy is designed for the

H.J. Biersack and P.H. Cox (eds), Nuclear Medicine in Gasteroenterology, 217–237

detection of small tumours typically under 2.5 cm and particularly under 1 cm diameter where the cancer cell is the dominant cell.

Clinical reasoning leads to the same conclusion, the requirement is for a non invasive technique to demonstrate the tumour, *not detectable by other modalities*, the subclinical subradiological recurrence, the viable peritoneal deposit that changes a 'complete remission' to a one year survival chance.

Radioimmunoscintigraphy is able to detect the tiny tumour because it gives a positive signal which depends not on tumour dimensions but on number of exposed antigens. Small tumours also have an adequate blood supply necessary for the delivery of the antibody. If a tumour the size of a pinhead had sufficient radioactivity upon it, it would be detectable externally. Thus the requirement for earlier detection of smaller tumour is to strive for the best radiolabel and the best antibody with the highest binding affinity and specificity for the tumor cell.

Radioimmunoscintigraphy is not designed to image the obvious nor is it designed to be all embracing. The detection of liver metastases may often be difficult with radioimmunoscintigraphy because of the natural uptake of antibodies by the reticuloendothelial tissue and liver metastases are reasonably well detected by ultrasound. Only a practitioner of 'nuclear medicine über alles' would deny its complementary role with other imaging techniques. Another example is the identification of a palpable or radiologically evident mass in the pelvis at follow up of a patient after rectal surgery. X-ray computed tomography may often be unable to distinguish post surgical fibrosis and repair from a viable recurrence. Radioimmunoscintigraphy shows no uptake in the former and high uptake in the latter.

The major contribution of radioimmunoscintigraphy in gastroenterology is in the field of colorectal cancer. It is effective but not particularly useful for primary diagnosis because endoscopy techniques and the barium enema are well established and more recently per rectal ultrasound is finding favour. Radioimmunoscintigraphy has its major role in the follow up and further management of patients after their primary surgery is completed. It is able to detect recurrent disease before serum markers are elevated and to localise disease when serum markers are elevated. It helps to identify recurrent disease when clinical and radiological indicators are suspicious and may lead to a more radical approach to the patients' management.

The antigen

Every cell surface contains a multitude of compounds which if injected into an appropriate recipient would elicit an immune response. These compounds are called antigens and the specific part of an antigen that reacts with the active centre of the antibody is called an 'epitope.' The prime question of radioimmunoscintigraphy is whether cancer specific antigens exist. At present no such antigens have been identified for colorectal cancer so that 'tumour associated' antigens are used instead. Nevertheless the oncogene – anti oncogene theory of cancer indicates that cancer specific oncoproteins will be identified. For the activation of particular DNA sequences in the gene in cancer means that complementary RNA sequences and hence amino acid sequences will give rise to oncoproteins against which cancer selective monoclonal antibodies may be made for use in radioimmunoscintigraphy. The principle has already been demonstrated by Chan and Sikora [1] in lung cancer for the C-myc oncogene.

In the present situation three classes of antigens are under investigation in colorectal cancer: the dedifferentiation antigens such as carcinoembryonic antigen, CEA; the tumour associated antigens such as TAG 72 where a homogenised carcinoma metastasis is used as the antigenic stimulus and the monoclonal antibody B72.3 is obtained in this way; and normal cell surface epithelial antigens protected by the basement membrane from access from the blood stream. The architectural description of the malignant process makes such antigens available in high density to react with injected monoclonal antibody for radioimmunoscintigraphy.

Carcinoembryonic antigen

CEA is a dedifferentiation antigen [2] expressed in almost all colorectal cancers, a large number of other cancers and normal tissues [3]. It is a shed antigen [4] evident in the colonic mucosa in the blood stream in the lymphatics and normal lymph nodes draining the tumour. There has been an extensive evaluation of CEA as a tissue and serum marker [3]. A number of non specific factors increase its release from normal cells such as cigarette smoking.

CEA has a number of epitopes leading to a range of monoclonal antibodies: one for the CEA epitope on granulocytes used for white cell labelling, many 'non-specific' cross reacting colonic antigens and a few more specific for colorectal cancer.

The use of radiolabelled anti-CEA polyclonal antibody for the detection of colorectal cancer was pioneered by Mach [5] and by Goldenberg [6]. Monoclonal antibodies against CEA for use in colorectal cancer have been selected so as to be free of non specific cross reactions such as C46 (Amersham) [7] or BW 431/26 (Behring) [8].

TAG 72 and B72.3

The antigen TAG 72 was developed using antigen membrane enriched cell abstracts of a liver metastases from human breast adenocarcinoma [9]. TAG 72 is expressed in several epithelial derived cancers including many but not all colonic adenocarcinomas and invasive ductal breast carcinoma, many epithelial ovarian cancers as well as pancreatic, gastric and oesophageal cancers. TAG 72 does not occur in tumours derived from neural tissue, bone marrow or fibrous stroma. It is not generally observed in adult normal tissues except secretory endometrium. It is present in some benign breast and colon tumours. As it is detected in foetal colon, stomach and oesophagus, it may also be called an oncofoetal antigen [10]. B72.3 is a monoclonal antibody against TAG 72 which has been extensively investigated in colorectal cancers [9–12].

Colonic crypt surface antigen and PR1A3

This antigen was discovered while investigating normal colonic epithelium for genetic markers [13]. It is situated in the epithelial surface lining of the columnar cells of the upper crypts of the colon. It is found marginally in the mucosa of the stomach and trachea. It is not found in other normal tissues.

It is protected from the circulation by the basement membrane of the colonic mucosa and it does not appear to be shed either into the blood stream, lymphatics or large intestine. Monoclonal Antibody PR1A3 against this antigen reacts with all types and stages of colorectal cancer and particularly well with undifferentiated colorectal cancer which is the most likely to metastasize.

The monoclonal antibody

A monoclonal antibody is produced from a hybridoma from cell fusion between spleen cells taken from an immunised mouse and cells from a mouse myeloma [14]. After selection of the hybridoma producing the 'best' antibody which is a lengthy and difficult procedure, it is grown in continuous cell culture and the monoclonal antibody is harvested at intervals. The production of monoclonal antibody for human use requires high pharmaceutical standards of quality assurance as set out in Begent's sensible guidelines [15]. Unfortunately, because a monoclonal antibody is a biological product, an over exaggerated perception of hazard, particularly of viral contamination, has lead to the formulation of a set of regulatory guidelines for conmercial manufacturers [16] that threatens to destroy the commercial viability of radioimmunoscintigraphy and radioimmunotherapy [17, 18].

The monoclonal antibody is a gamma globulin which may exist in a number of forms or classes. For radioimmunoscintigraphy typically class IgG1 or IgG2a monoclonal antibodies are used. The gamma globulin consists of two light chains and two heavy chains held together by S–S bridges in the form of a Y. The active centre is at the end of each light – heavy chain combination at the end of the arms of the Y known as the 'hyper variable' region which is supported by a 'frame.' The base of the Y made of heavy chain only is the 'Fc' portion which binds to reticuloendothelial tissue. Proteose digestion of the whole gamma globulin may give either a Fab fragment – one arm of the Y, or an F(ab')2 fragment which is both arms of the Y with a small part of the base and having both active centres. This reduces the nonspecific uptake and reduces the total size of the molecule making it more diffusible into tumour. However, the F(ab')2 fragment is cleared more rapidly from the blood and has a more extensive tissue distribution than whole antibody. It involves a further preparative step, may be less easy to radiolabel and may have a lower binding affinity for the tumour than whole antibody.

The immunoreactivity of the antibody and fragment is crucial. An antibody antigen binding constant of 10^{-9} mol/l^{-1} or more is required. It appears that circulating antigen such as CEA has a much lower binding affinity than the antigen fixed on tissue, so a high circulating CEA does not impede the uptake of anti CEA by colorectal cancer. This immunoreactivity must be retested after the radiolabelling procedure by direct radioimmunoassay or an ELISA technique. However, quantitative evaluation of immunoreactivity is not straightforward.

After intravenous injection into a patient, the initial rate of uptake is the greatest. The fractional uptake of the amount administered decreases with time as the concentration in the blood supplying the tumour decreases with time. The arrival of the antibody in the vicinity of the antigenic binding sites on the cells depends on physical factors: capillary permeability and the effects of the chemical composition and charge of the intercellular fluid and cellular environment on the diffusion and convection of the antibody. The higher the avidity of the antibody for the antigen, the greater the probability of an interaction and combination on the cell. For a typical whole antibody and a typical compact, reasonably vascular tumour, noting that tumour vessels are less responsive to autonomic and chemical control factors and tumour capillaries are more permeable than normal, about 75% of the total tumour uptake of antibody would occur in the first 12 h. For some antibodies uptake will be more rapid as shown by Buraggi [19] for antimelanoma antibody and for others particularly if they are relatively avascular the uptake rate will be slower. However, in gastrointestinal tumours adequate uptake will have occured within 24 h of injection. The old attitude that one should wait a week for uptake to be sufficient for imaging has been shown to be inappropriate [20, 21].

The radiolabels

The choice of radiolabel is crucial. The major determinant of a suitable radionuclide is its half-life of decay. The shorter the half-life, the greater the activity that may be administered and the higher the count rate obtained, given an upper acceptable limit of the absorbed dose of radiation that is permitted for diagnostic nuclear medicine. Ideally the physical half-life of decay of the radionuclide should be matched with information on the rate of antibody uptake by the target: rapid uptake favouring ^{99m}Tc (6 h) and ^{123}I (13.2 h), slower uptake favouring ^{111}In (67 h). ^{131}I is no longer appropriate

Table 1. Tumor detection efficiency.

	Dose MBq	Ratio to ^{131}I
^{99m}Tc	600	12.9
^{123}I	120	5.0
^{111}In	120	3.8
^{131}I	60	1.0

for diagnostic work giving Beta radiation and having an unnecessarily long half-life (8 day) and an inefficient gamma ray energy for the modern gamma camera.

The second essential requirement for high resolution tumour uptake and for high sensitivity for subclinical and subradiological tumour detection is to obtain the highest achievable count rate from the target by the imaging system.

An appreciation of the need for high sensitivity requires an understanding of the sources of signal degradation, which is usually summarized by the potentially misleading term, the signal to noise ratio. Noise is a term used generally for anything that degrades the signal and has many sources but the most important is the signal itself. The weaker the signal, the more the noise inherent in the signal itself overwhelms the effect of any contribution from non-target background.

This inherent noise arises because the radioactive decay of a radionuclide with the emission of gamma rays is a random process governed by Poisson statistics. Poisson noise depends on the square root of the count rate so that the lower the count rate, the disproportionately higher is the noise. Thus at 10000 counts per minute (cpm) the noise is 100 cpm – that is, 1% of the counts recorded – whereas at a count rate of 100 cpm the noise is 10 cpm – that is 10% of the counts recorded. Low count rates give 'noisy' images where the target is less separable from background and the high variability of the background itself may give 'noise blobs' which may be falsely interpreted as target signals. A good signal enables detection however high the non-specific background activity [20, 21]. Putting together the count rate with the kinetics of antibody uptake and the effective half-life of the labelled antibody means that given a tumour of a particular size, the higher the count rate is and the more rapid the antibody uptake, the earlier is its detection. A comparison of the radionuclide for tumour detection efficiency with a signal to noise ratio of 1.25:1 is shown in Table 1 (from [20]).

Thus the high count rate signal from a ^{99m}Tc labelled antibody is twice as good as that from ^{123}I and three times as good as that from ^{111}In at detecting tumour on theoretical grounds. The multicentre study of radiolabelled anti-melanoma antibody showed superior results with ^{99m}Tc as the label over ^{111}In confirming this approach [22].

Once the signal has been maximised, approaches to the reduction of tissue background activity (which is quite different from that due to the noise inherent in the signal itself) can be addressed. Single photon emission tomography is a useful way of separating activity in normal structures such as the bladder (^{99m}Tc or ^{123}I label) or bone marrow (^{111}In label) from sites of specific uptake. Reconstruction of the data in transverse, coronal and sagittal sections gives a three dimensional view of the tumour uptake in relation to its environment. A high standard of camera quality control and technique is essential [23].

Tissue background subtraction techniques have progressed from a double radionuclide approach (^{131}I and ^{99m}Tc [6]) to the subtraction of an early image from a later one [24] to kinetic analysis with probability mapping [25].

These techniques are *not* appropriate for abdominal gastrointestinal cancer if there is movement of radiolabel in the small or large bowel as is typical for anti CEA antibody and for antibody labelled with ^{111}In or radioiodine. These approaches may be applicable to the search for liver metastases.

Radiolabelling of monoclonal antibodies

This is outside the scope of this chapter. Radiolabelling with ^{123}I is usually by the iodogen technique as used for the human milk fat globule HMFG2 antibody [26].

Radiolabelling with ^{111}In is usually by the bifunctional chelate method using the cyclic bis anhydride of DTPA [27, 28].

Radiolabelling with ^{99m}Tc has revolutionised radioimmunoscintigraphy. The Schwarz technique [29] provides a simple method which is applicable in every nuclear medicine department. The S–S bonds holding the heavy chains together in the stem of the Y of the gamma globulin are opened using 2-mercapto ethanol. The antibody may then be stored frozen. At the time of use, some tin as a reducing agent is supplied by a bone imaging kit followed by 600 MBq ^{99m}Tc pertechnetate eluate. The resulting ^{99m}Tc labelled antibody is stable in vivo with no thyroid uptake of ^{99m}Tc even after 24 h, the patient having received no thyroid blocking medication. The technique as modified by Mather is given in Table 2 and has been successfully used for four different antibodies in clinical use to date.

Imaging technique

Intravenous injection of between 0.5 mg and 2 mg of monoclonal antibody is sufficient. Amounts over 30 mg of antibody infused will slightly reduce nonspecific liver uptake but are not worthwhile. Before injection it is *essential* to ask for a history of allergy from the patient, with allergy to foreign protein being an absolute contraindication to the study. An intradermal skin test is not advised both because a negative test does not exclude a reaction and because intradermal injection is particularly effective at sensitising the patient and initiating a human antimouse antibody, HAMA, response. Vital signs should be monitored once before and after injection and the patient observed for 30 min. Since an image taken 10 min after injection is an essential baseline for interpreting subsequent images, this observation period is straight forward.

Imaging with indium-111 labelled antibody

For imaging, the patient lies supine on the couch with a gamma camera set over the pelvis. The camera is used with a medium energy (up to 300 KeV) general purpose parallel hole collimator and set up for the two photo peaks 171 and 247 keV of ^{111}In-Indium with 15% and 20% windows respectively. Uniformity and linearity quality control must be performed regularly. The data is transferred on line to a computer for subsequent analysis.

The contents of the vial containing the ^{111}In-labelled anti-CEA in a measured concentration are transferred to a weighed syringe and needle and assayed in an ionizing chamber. After injection the syringe and needle are reweighed and residual activity reassessed so that the activity of the ^{111}In-Indium label and the weight of antibody injected are known. Injection of between 3 and 4.5 mCi, (115–165 MBq) of ^{111}In-labelled anti-CEA or PR1A3 in a volume of 3–5 ml intravenously into an antecubital vein is performed slowly over 10 s. Anterior and posterior static images of the lower chest and upper abdomen, and lower abdomen and pelvis are made at 10 min, and 24 h. Single photon emission tomography, SPECT, with 360° rotation with 64 projections taking 32 min is performed in most patients, at 24 h. Further planar images are made at daily intervals until and including the morning of operation if performed, or for three days. Each planar image should contain 800000 counts. For each planar image, a similar image with radioactive markers over six prominent bony landmarks, symphysis, iliac spines, costal margins and xiphisternum, are taken to facilitate the subsequent repositioning of the patient [7].

The planar and single photon emission tomography images should be analysed in the absence of clinical and radiological information and decisions made as to whether an unequivocal primary tumour or recurrence is seen, and whether the liver is involved. Sites of equivocal uptake should be recorded as well as comments on image quality, on the degree of general liver, marrow, vascular, renal and colonic uptake and whether this interfered with the assessment. After signed reports have been made both of the diagnosis and of the image acceptability, these may be compared directly with the presentation and clinical findings, the diagnosis made at the surgery and the histopathological findings (Figs 1–5).

Surgical specimens may be analysed to evaluate the degree and specificity of tumour uptake.

At surgery for primary tumours, the tumour bearing colon or rectum is excised with its mesentery and sent to the histopathologist for pinning and

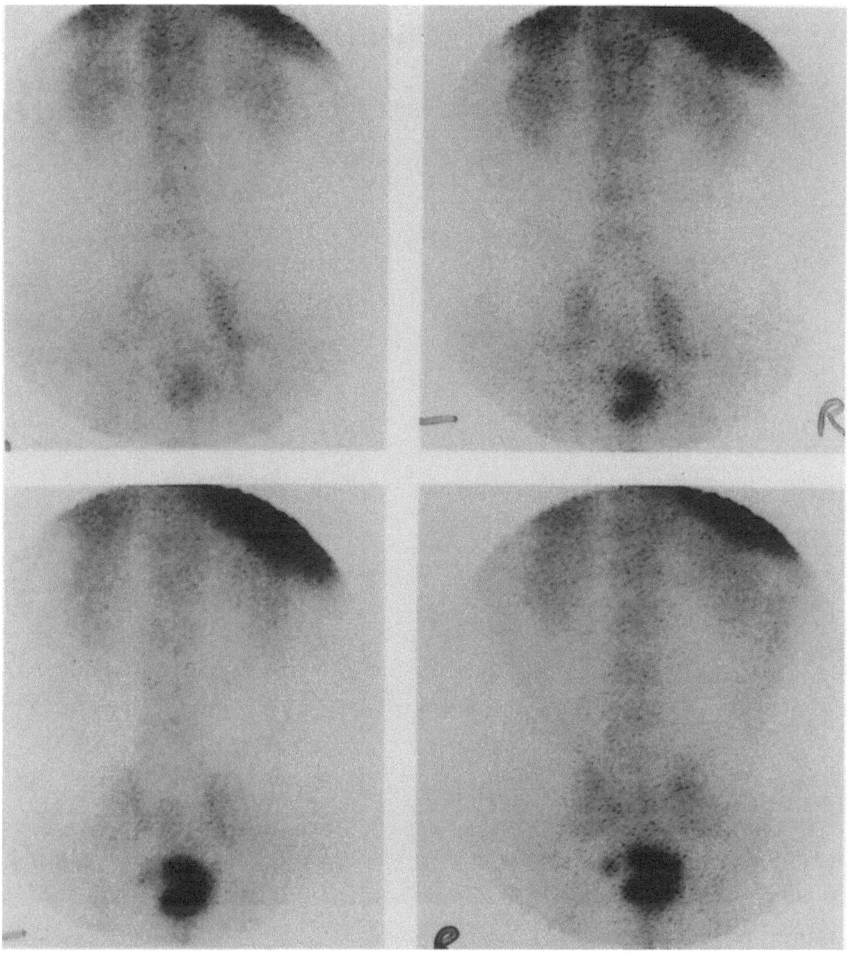

Figure 1a. ^{111}In-antiCEA in colorectal cancer. Posterior views of abdomen at 10 min top left; 4 h, top right; 24 h, bottom left; 48 h bottom right. Increasing uptake with time is noted in a tumour in the pelvis with uptake in a lymph node visible at 48 hrs to the left of the tumour. (Courtesy International Journal of Colorectal Disease).

fixing. The sites of lymph nodes, polyps, adenoma, carcinoma, site of previous surgery and the presence of other disease are recorded and a map of the specimen drawn. Staging is then performed using the Dukes' classification [30] (Fig. 4d, e).

Dukes A: the tumour has spread into the tissue of the bowel wall but not beyond the muscularis propria;

Dukes B: where the spread is beyond into peri-rectal or peri-colonic tissue but no lymph nodes are involved;

Dukes C: where lymph nodes are involved in addition.

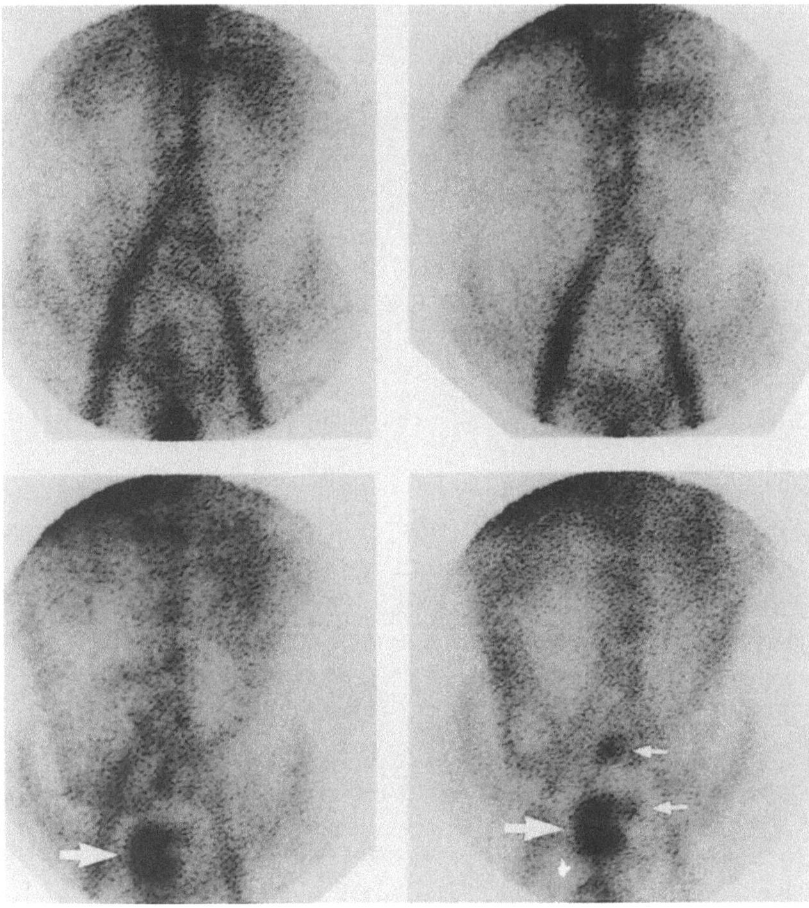

Figure 1b. Anterior views of abdomen. Image times as for 1a. Increasing uptake with time is noted in a tumour of the pelvis (large arrow) with lymph node uptake (small arrows). (Courtesy International Journal of Colorectal Disease).

The specimen is then imaged under the gamma camera (Figs 1c, 3b, 4c). It is then returned to the histopathologist to dissect out the tumour and lymph nodes. Samples of these and of the mucosa are obtained and weighed. These specimens are then counted in a sample counter set up for ^{111}In-Indium with appropriate background tubes and standards so that the activity in each specimen can be determined. These values are decay corrected for the time since injection and expressed as a percentage of the injected dose typically as 10–2% per gram of specimen.

The presence of CEA or the antigen for PR1A3 may be also confirmed by immunoperoxide staining on the specimens.

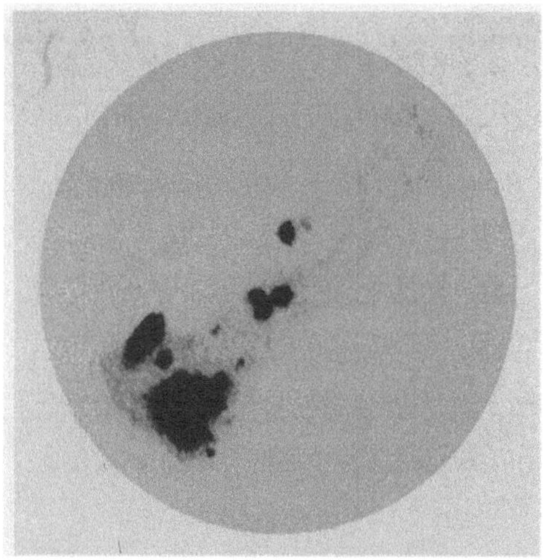

Figure 1c. Surgical specimen imaged with gamma camera. The uptake in the tumour mass is seen to the left. Tumour was moderately well differentiated with 1.4×10^{-2}% injected dose per gram, weight 31.4 g. Tumour to mucosa ratio 7:1. Uptake in many lymph nodes is seen but *none* were involved with malignant cells on histology. Dukes B. (Courtesy International Journal of Colorectal Disease).

Imaging with ^{99m}Tc labelled antibody

The gamma camera is prepared with a low energy (up to 200 keV) general purpose parallel hole collimator and set up for the 140 keV photopeak with a 15% window and with data transfer on line to the computer.

Following the same patient preparation as above the ^{99m}Tc labelled antibody anti CEA [8] or PR1A3 is injected intravenously [29]. Images are taken

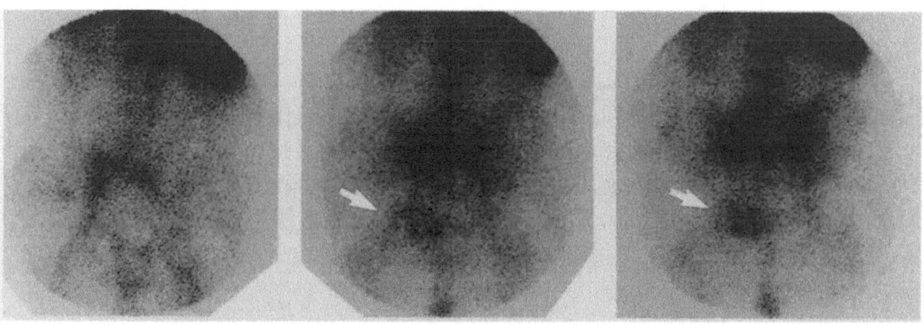

Figure 2. ^{111}In-antiCEA in recurrence of carcinoma of the rectum. Posterior view of pelvis at 10 min, 24 h and 48 h – left to right. Increasing uptake is seen in a lesion to the left of the midline in the pelvis (arrowed). Normal marrow, renal and liver activity is also seen. This image confirmed the presence of viable tumour at the site of a possible abnormality on X-ray C.T. (Courtesy International Journal of Colorectal Disease).

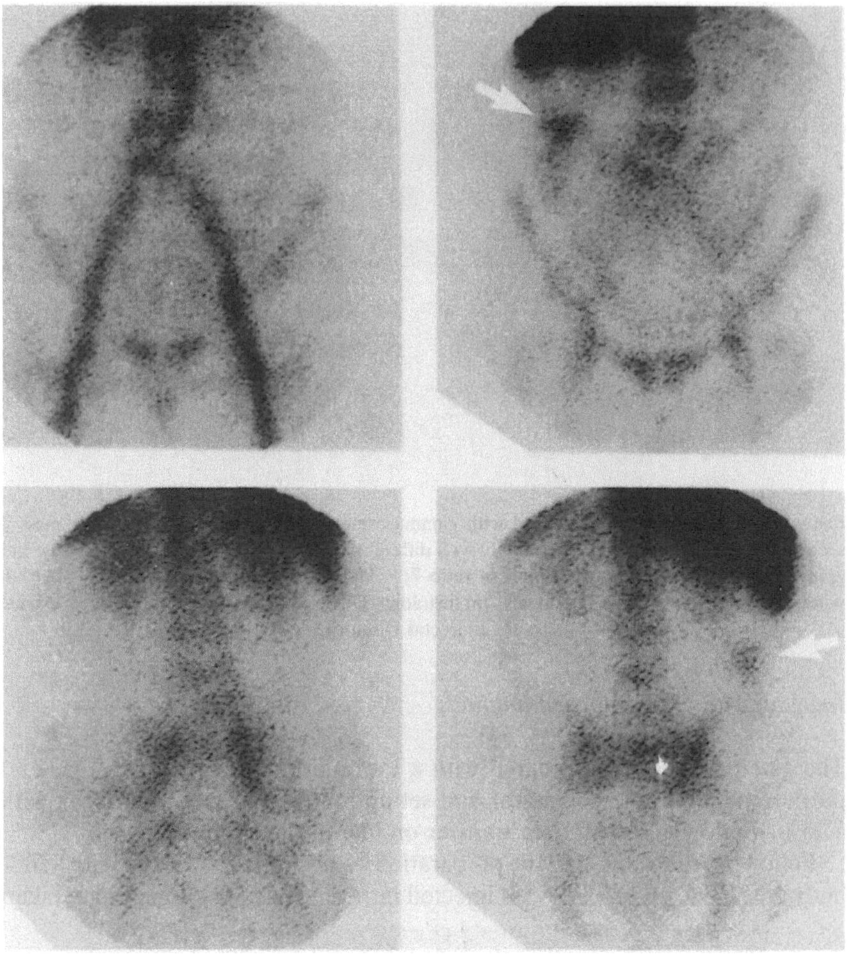

Figure 3a.

at 10 min, 3, 6, and 24 h. Each planar image contains 800000 counts. An image of the thyroid is taken at 30 min and at 24 h to confirm the in vivo stability of the labelling. No thyroid blockade is required (Figs 6, 7).

Clinical applications

The major clinical application of radioimmunoscintigraphy in gastroenterology is in the management of colorectal cancer after primary surgery. The Dukes' classification has been related to prognosis Dukes A 95%, Dukes B 60%, Dukes C 30% five year survival. Dukes C is the commonest stage found at primary surgery and has a 50% chance of recurrence at one year. Conventionally serial serum CEA is used to follow up such patients, but the

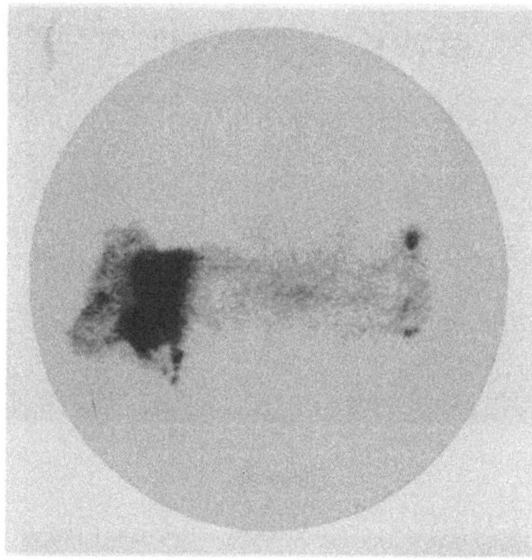

Figure 3b.

Figure 3. ^{111}In antiCEA study of Carcinoma of the Ascending Colon. (a) Top, anterior view of the abdomen at 10 min, and 24 h (left to right). Bottom: posterior view at the same time. Increasing uptake is seen in a lesion in the ascending colon in the region of the hepatic flexure (arrowed). Marrow and liver uptake is also seen, but no bladder activity. (b) Image of the surgical specimen. Uptake in the tumour is seen to the left and in a polyp on the right. Tumour was poorly differentiated with only 0.046×10^{-2}% injected dose per gram, weight 16.75 g. Tumour to mucosa ratio 2.4:1. Lymph node uptake is less evident than in Fig. 1b, 6 of 13 nodes were involved, Dukes C. (Courtesy International Journal of Colorectal Disease).

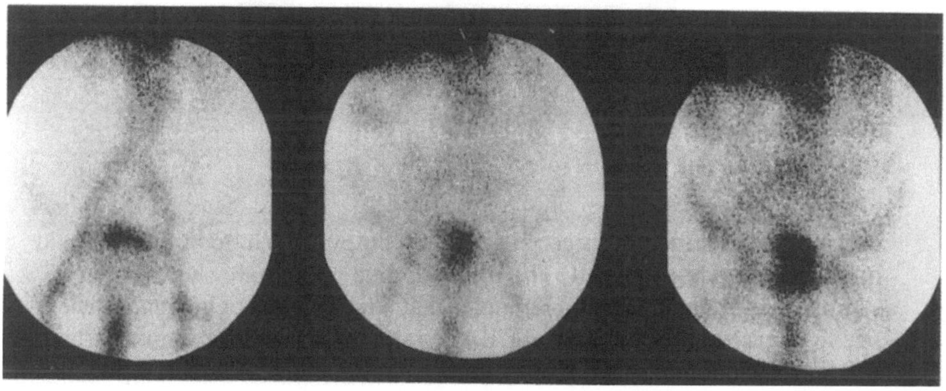

Figure 4a.

nonspecific rises and falls of serum CEA require that there is an increasing serum level during three successful months. We have started to perform radioimmunoscintigraphy at the one year routine follow up appointment after primary surgery and Dukes C staging. Already two of five such patients have demonstrated clearly recurrent disease at a time when their serum CEA was within the normal range. Such disease was confirmed and lead to the

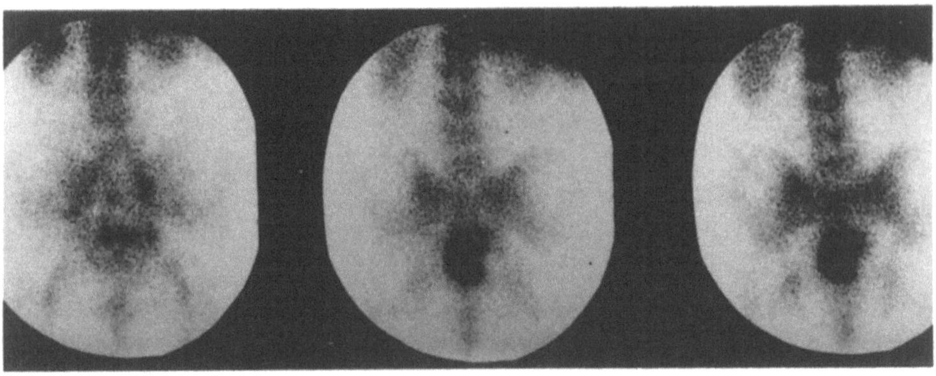

Figure 4b.

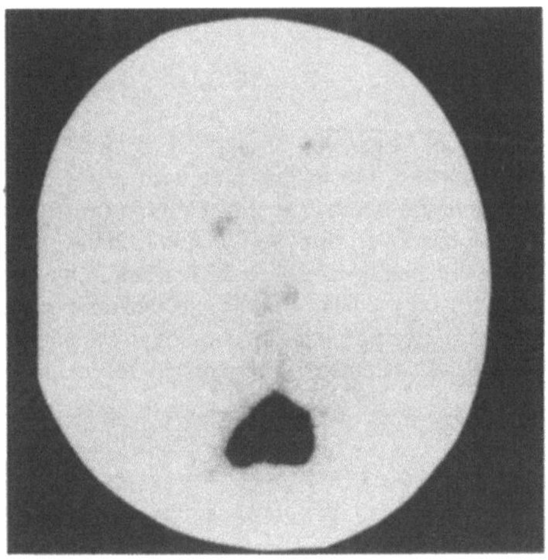

Figure 4c.

initiation of treatment – surgery and radiotherapy, which opportunity would otherwise have been missed. The results in one patient are shown in Fig. 7.

This type of finding gives a reminder that a tumour has to be of a sufficient size to release a sufficient load of CEA to raise the serum level above normal, whereas radioimmunoscintigraphy gives a direct evaluation of the presence of a recurrence as well as locating its sites. The sensitivity of radioimmunoscintigraphy with Indium-111 labelled anti-CEA is 95% in our hands in a prospective study of primary and recurrent cancer [7].

In comparison with X-ray CT Indium-111 labelled anti-CEA had a slightly greater sensitivity than X-ray CT and a greater specificity (Granowska, unpublished).

X-ray CT was able to demonstrate an abnormal mass in the pelvis but not whether it was a viable tumour or post surgical fibrosis. The X-ray CT also

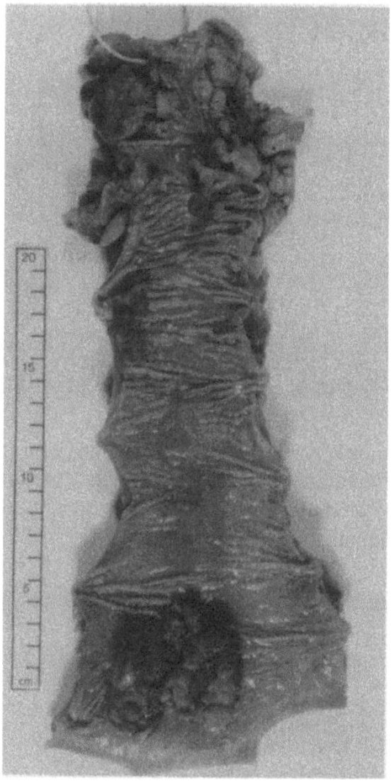

Figure 4d.

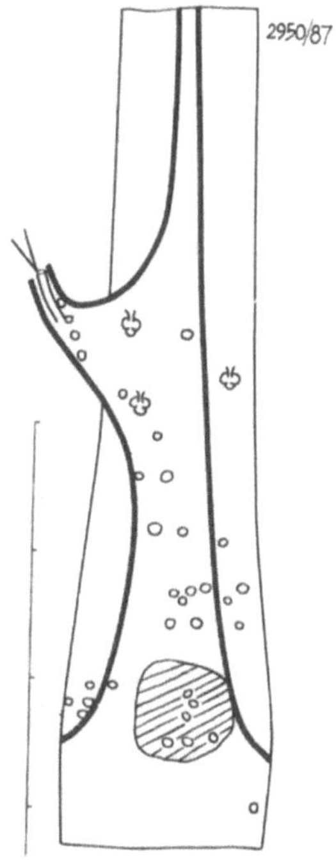

Figure 4e.

Figure 4. ^{111}In PR1A3 study of carcinoma of the rectum. (a) Anterior view of the abdomen at 10 min, 24 h and 72 h (left to right). (b) Posterior view of the abdomen at the same times. Both show increasing uptake with time in tumour. (c) Image of the surgical specimen showing high tumour uptake below and minor uptake in three small polyps. (d) The pathological specimen. Moderately differentiated adenocarcinoma containing 0.67% injected dose per gram, tumour weight 7.2 g with a tumour to mucosa ratio of 46:1. (e) The histopathologists diagram. Note the polyps and that all nodes are not involved, Dukes A. Note that the image 4c shows no uptake in these normal nodes in contrast to Fig 1c.

missed plaque like recurrences on the pelvic or abdominal wall as also demonstrated by Chatal [31].

Once the serum CEA is elevated then radioimmunoscintigraphy with anti-CEA is the technique of choice to locate the site of the tumour. Often at this time conventional investigations are negative.

High sensitivity is not only due to the choice of antibody but also due to great care in the performance of the imaging technique and in interpreting the results. The 10 min image is an essential base line, a template against which other images may be compared. It should be noted that specific uptake increases with time over 24 h whereas nonspecific uptake after an initial distribution phase decreases with time. However, with anti-CEA there are a number of variants. The high mucosal secretion of CEA means that high

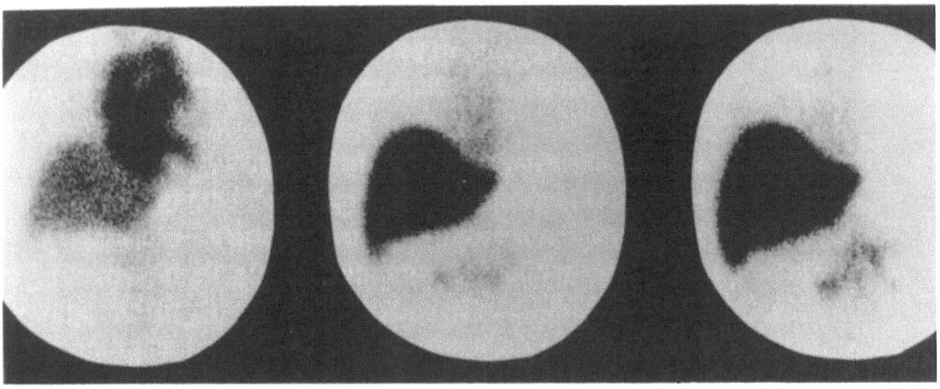

Figure 5. ^{111}In PR1A3 study of a transverse colon tumour. Anterior views at 10 min, 24 h 72 h (left to right). This tumour was a moderately differentiated adenocarcinoma, weight 12.6 g with 0.18%, injected dose per gram, tumour to mucosa ratio 4.6 Dukes B. Note the high liver uptake relative to the multifocal uptake in the tumour. Compare with Fig 6.

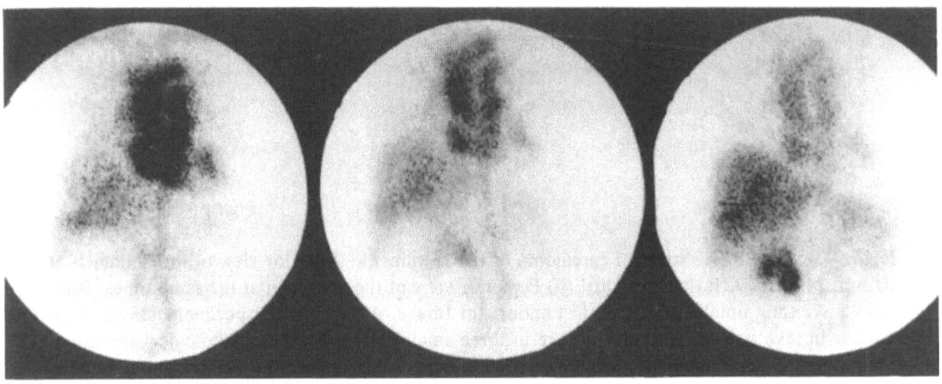

Figure 6. ^{99m}Tc PR1A3 study of transverse colon tumour. Anterior views at 5 min, 6 h and 24 h. This tumour was a poorly differentiated mucous adenocarcinoma, 7.8 g, with 0.014% of the injected dose per gram, tumour to mucosa ratio 63 : 1, Dukes B. Note that the tumour uptake increases with time and is relatively greater than the liver uptake, compare Fig 5. This is due to the increased renal uptake, but no thyroid uptake was seen.

large bowel activity is usual but the bowel activity moves with time whereas tumour activity does not. Several inflammatory bowel diseases however secrete CEA and particularly in the case of Crohn's disease high fixed uptake unwavering with time may be seen and must be remembered in the differential diagnosis. Adenoma and polyps often take up anti-CEA. However, many consider these to be premalignant particularly when over 3 cm diameter and require surgery in their own right [32]. Is malignancy or the need for operation the gold standard with which to compare the results of radioimmunoscintigraphy when specificity is the consideration?

One important problem with anti-CEA is that, although CEA is a dediffer-

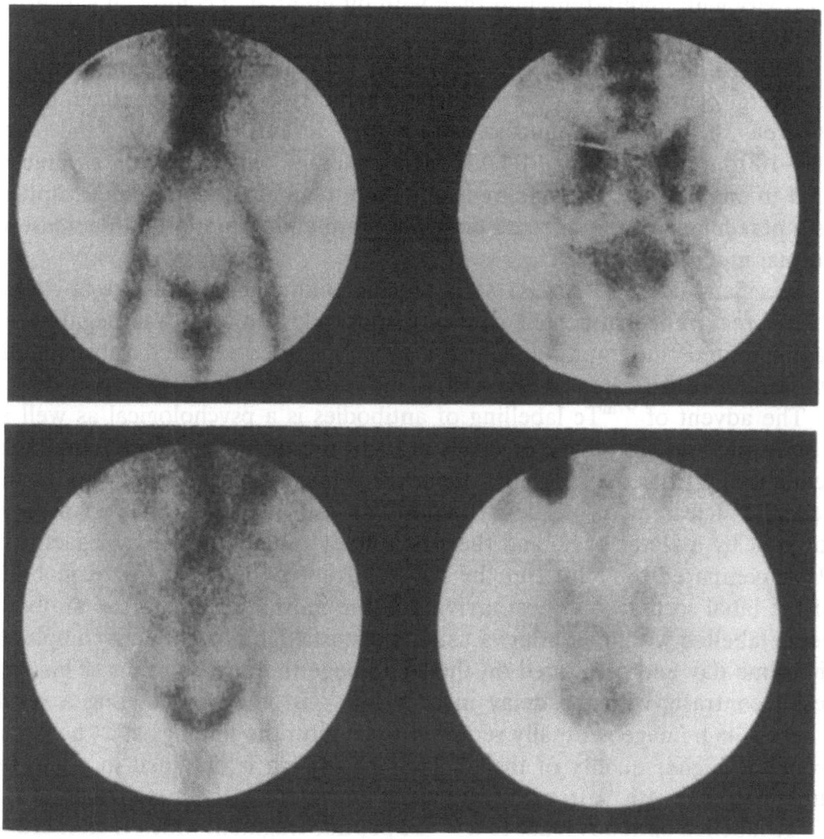

Figure 7. ^{99m}Tc PR1A3 study of recurrent adenocarcinoma of the pelvis. (Left) Anterior views of the abdomen 10 min, top; 24 h, bottom. (Right) Posterior views of the abdomen 6 h top, 24 h bottom. Note the uptake in the pelvic mass posteriorly with the anteriorly placed bladder U shaped beneath it, and the high renal uptake. This patient had an undifferentiated rectal carcinoma Dukes C excised 1 year previously and was symptom free with a normal serum CEA when this recurrence was found by routine radioimmunoscintigraphy. At 1 year Dukes C have about a 50% chance of recurrent disease.

entiation antigen, it is poorly produced by undifferentiated tumour, uptake being 12.9 times lower than on average in moderately differentiated tumour in our study.

The importance of this observation is that the poorly differentiated tumour which is more likely to spread and metastasise is likely to be less easy to detect than the well differentiated tumour. In our study 80% of poorly differentiated tumours were staged as Dukes C whereas 93% of the moderately or well differentiated tumours were staged as Dukes A or B.

This problem has been met by the use of the new antibody for colorectal cancer called PR1A3 and described on page 220. This has a much higher reactivity with undifferentiated than with differentiated colorectal cancer up to 20 times on average [33].

Both with ^{111}In-PR1A3 and now with ^{99m}Tc PR1A3 the sensitivity for colorectal cancer was 92% and including the two villous adenoma 100%. Whereas the tumour to mucosa ratio with ^{111}In anti-CEA was of the order of 4–10 to one, that with PR1A3 was an average twice this with a value of 63:1 in one (Fig 6). This finding is due both to an increase in tumour uptake as a percentage of the injected dose per gram and an important decrease in normal mucosal uptake.

Specificity using ^{99m}Tc PR1A3 is also high although patients are few. One with a recurrence suspected clinically and endoscopically was negative on radioimmunoscintigraphy and histology of the area resected subsequently showed no recurrence.

The advent of ^{99m}Tc labelling of antibodies is a psychological as well as a scientific breakthrough, for it has brought the technique of radioimmunoscintigraphy from the specialist department into routine applicability in the general nuclear medicine department. The use of the ^{99m}Tc label reduces the cost by a factor of 80 and the whole body radiation dose by a factor of 20 as compared to ^{111}In. But the major advantage is the ready availability of the label so that a request arriving in the morning leads to the antibody being labelled within an hour, a usual diagnostic 6 h image being completed the same day and confirmed on the 24 h image the next morning. This 24 h result contrasts with the delay in ordering ^{111}In and in obtaining a result where a 48 h image is usually required to confirm the findings at 24 h. Apart from the higher quality of the ^{99m}Tc image which is obtained in a shorter time than that with ^{111}In with less movement artefact, the metabolism of the ^{99m}Tc labelled antibody leads to a ^{99m}Tc peptide fragment which is glomerular filtered by the kidney, reabsorbed at the proximal tubule and further metabolised to deposit ^{99m}Tc as with many other metals into the proximal tubular cells. This reduces the ^{99m}Tc antibody available for liver uptake. As a result the intensity of liver uptake is much less than that with ^{111}I, so that, as Baum [8] has shown many liver metastases can be identified as focal areas of increased uptake against a moderate normal liver background.

Other methods of ^{99m}Tc labelling of antibodies are being developed and applied to colorectal cancer [34] and the future approach of genetically engineered humanised 'chimeric' antibodies [35] where only the hypervariable reactive region is of murine origin is the next advance. Even genetically engineered metallothionein to incorporate the Technetium 99m label is being linked to the humanised antibody [36]. Such an approach should remove much of the nonspecific reticuloendothelial system uptake and reduce the biochemical HAMA response. The reaction rate in diagnostic clinical practice is already under 1 per 1000, about the same level as that of a renal ^{99m}Tc DTPA study in spite of the HAMA response.

Thus the future combination of the ^{99m}Tc label, more cancer specific antibodies or genetically engineered antibody-like reagents and a greater appreciation of the intentions of radioimmunoscintigraphy should make the technique an indispensible part of gastroenterological medical and surgical practice.

Radioimmunoscintigraphy of other cancers

An increasing number of monoclonal antibodies have been tested or are undergoing evaluation in the whole range of gastrointestinal cancers. The principles and practice are generally similar to those for colorectal cancer which can be taken as a role model. CEA is produced by stomach and some pancreatic cancers. B72.3 reacts with stomach and some pancreatic cancers. Anti CA 19–9 reacts with both stomach, pancreatic and colorectal cancers. However, none of these techniques has approached the development and usage of radioimmunoscintigraphy seen in colorectal cancers.

The approach to carcinoid tumours and pancreatic tumours may be better made through the use of receptor binding radiopharmaceutical. These include, ^{131}I and ^{123}I MIBG, metalodobenzyl-guanidine which reacts with some carcinoids [37], and ^{123}I octreotide which binds to somatostatin receptors in pancreatic tumours and many-gastrointestinal endocrine tumours such as insulinoma and vipoma [38].

Conclusions

Radioimmunoscintigraphy is a new technique for demonstrating and localising cancer, particularly recurrences which is already aiding and altering the management of patients with colorectal cancer. A pathway towards cancer specific diagnosis has been aided by the development of ^{99m}Tc labelling of antibodies which retain stability in vivo. This has broken the psychological and practical logistic barriers and now enables radioimmunoscintigraphy to become practised routinely in a general department of nuclear medicine.

Acknowledgements

We gratefully acknowledge the support of the Imperial Cancer Research Fund and the use of the facilities of St. Bartholomew's Research Centre.

References

1. Chan S, Sikora K (1987) 'The potential of oncogene products as tumor markers.' *Cancer Surveys* 185–207
2. Gold P, Freedman SO (1965) 'Specific carcinoembryonic antigens of the human digestive system.' *J Exp Med* 122: 467–481.

3. Go VLW (1976) 'Carcinoembryonic antigen, clinical application.' *Cancer* 37: 562–566
4. Moshakis V, Omerod MG, Westwood JN et al. (1982) 'The site of binding of anti CEA antibodies to CEA *in vivo*: An immunocytochemical and autoradiographic approach.' *Brit J Cancer* 46: 18–21.
5. Mach JP, Carrel S, Forni M et al. (1980) 'Tumour investigation of radiolabelled antibodies against carcinoembryonic antigen in patients with carcinoma.' *New Eng Med* 303: 5–10.
6. Goldenberg DM, Kim EE, Deland FH et al. (1980) 'Radioimmunodetection of cancer with radioactive antibodies to carcino embryonic antigen.' *Cancer Res* 40: 2984–2992.
7. Granowska M, Jass JR, Britton KE, Northover JMA (1989) 'A prospective study of the use of ^{111}In labelled monoclonal antibody against carcino embryonic antigen in colorectal cancer and of some biological factors affecting its uptake.' *Int J Colorect Dis* 4: 97–108.
8. Baum RP, Hertel A, Lorenz M et al. (1989) '^{99m}Tc labelled anti-CEA monoclonal antibody for tumour immunoscintigraphy: First clinical results.' *Nucl Med Commun* 10: 345–352.
9. Colcher O, Han P, Nuti M, Schlom J (1981) 'A spectrum of monoclonal antibodies reactive with human mammary tumour cells.' *Proc Natl Acad Sc USA* 78: 3199–3203.
10. Schlom J, Colcher D, Roselli M et al. (1989) 'Tumour targetting with monoclonal antibody B72.3.' *Nucl Med Biol* 16: 137–142
11. Esteban JM, Colcher D, Sugerbaker P et al. (1987) 'Quantitative and qualitative aspects of radiolocalization in colon cancer patients of intravenously administered Mab B72.3.' *Int J Cancer* 39: 50–59
12. Salvatore M, Lastoria S, Del Vecchio S, Mansi L (19??) 'Detection of Colon Cancer with radiolabelled monoclonal antibodies. *Nucl Med Biol* 16: 103–104.
13. Richman PI, Bodmer WF (1987) 'Monoclonal antibodies to human colorectal epithelium: markers for differentiation and tumour characterisation.' *Int J Cancer* 39: 317–328.
14. Kohler G, Milstein C (1975) 'Continuous culture of fused cells secreting antibody of predefined specificity.' *Nature* 256: 495–6.
15. Begent RHJ (1986) 'Working party on the clinical use of antibodies: operations manual.' *Brit J Cancer* 54: 557–568
16. Committee for Proprietary Medical Products Ad Hoc working party of Biotechnology/Pharmacy. Notes to applicants for Marketing Authorizations on the production and quality control of monoclonal antibodies of Murine origin intended for use in man. Commission of the European Communities III/859/86–EN Rev. 7. Final 1987.
17. Britton KE (1989) 'Potential clinical applications of monoclonal antibodies in bringing biotechnology to the market.' *British Institute of Regulatory Affairs*. In press.
18. Prentice T (1989) 'Cancer Research: Bureaucrats 'hinder search for cure.' *Times* Aug. 30th.
19. Buraggi GL, Turrin A, Cascinelli N et al. (1985) 'Immunoscintigraphy with antimelanoma monoclonal antibodies.' In: L. Donato & K.E. Britton (eds.), *Immunoscintigraphy*, pp. 215–253. London: Gordon & Breach.
20. Britton KE, Granowska M (1987) 'Experience with Iodine-123 labelled antibodies for imaging and therapy.' Scrivastava SC (ed) *NATO AS1 Series*, pp 177–191. New York: Plenum Press.
21. Britton KE, Granowska M (1987) 'Radioimmunoscintigraphy in tumour identification cancer surveys.' G: 247–267.
22. Siccardi AG, Buraggi GL, Callegaro L et al. (1986) 'Multicentre study of immunoscintigraphy with radiolabelled monoclonal antibodies in patients with melanoma.' *Cancer Res.* 46: 4817–4822.
23. Bares RE (1989) 'A brief guide to the practice of Radioimmunoscintigraphy and Radioimmunotherapy in Cancer,' in Britton KE, Buraggi G (eds), *Int J Biol Markers* 4: 106–118.
24. Granowska M, Britton KE, Shepherd J (1983) 'The detection of ovarian cancer using 123-1 monoclonal antibody.' *Radio Biol Radiother* (*Berlin*) 25: 153–160.
25. Granowska M, Nimmon CC, Britton KE, Mather SJ, Crowther M, Slevin ML, Shepherd

JH (1988) 'Kinetic analysis and probability mapping applied to the detection of ovarian cancer by Radioimmunoscintigraphy.' *J Nucl Med* 29: 594–607.

26. Epenetos AA, Britton KE, Mather S et al. (1982) 'Targeting of Iodine-123 labelled tumour associated monoclonal antibodies to ovarian, breast and gastrointestinal tumours.' *Lancet* ii: 999–1004.
27. Hnatovitch DH, Layne WW, Childs RL et al. (1983) 'Radioactive labelling of antibody: a simple and efficient method.' *Science* 220: 613–615.
28. Mather SJ (1987) 'Labelling with Indium-Ill. Safety and Efficacy of Radiopharmaceuticals,' in Kristensen K, Norbygaard E (ed), pp 51–56. Dordrecht: Martinus Nijhoff.
29. Schwarz A, Steinstraesser A (1987) 'A novel approach to ^{99m}Tc labelled monoclonal antibodies.' *J Nucl Med* 28: 721 (Abs).
30. Dukes CE (1932) 'The classification of cancer of the rectum. *J Path Bact* 35: 323–332.
31. Chatal JF, Saccavini JC, Furnoleau P et al. (1984) 'Immunoscintigraphy of colon carcinoma.' *J Nucl Med* 25: 307–314.
32. Day DW, Morson BC (1978) 'The adenoma carcinoma sequence. In the pathogenesis of colorectal cancer.' in Morson BC (ed), pp 48–61. Philadelphia: WB Saunders & Co.
33. Granowska M, Mather SJ, Britton KE et al (1989) 'A ^{99m}Tc labelled monoclonal antibody PR1A3 for Radiommunoscintigraphy, RIS, of colorectal cancer.' *J Nucl Med* 30: 748 (Abs).
34. Goldenberg DM, Ford EH, Lee RE et al. (1989) 'Initial clinical imaging results with a new ^{99m}Tc antibody labelling method.' *J Nucl Med* 30: 809 (Abs.)
35. Bischof-Delaloye A, Delaloye B, Buchegger F et al. (1989) 'Chimeric mouse-human anti CEA antibody of IgG4 isotype used in a pilot immunoscintigraphy study of patients with colorectal carcinomas.' *J Nucl Med* 30: 809 (Abs.)
36. Epps LA, Sun L, Arevalo M et al. (1989) 'Technetium (^{99m}Tc) labelled genetically engineered chimeric 17–1A G4K/Metallothionein antibody.' *J Nucl Med* 30: 794 (Abs).
37. Jodrell DI, Irvine AT, McCready VR et al. (1989) 'The use of 131–I-MIBG in the imaging of metastatic carcinoid tumours.' *Br J Cancer* 58: 663–664
38. Krenning EP, Bakker WH, Breeman WAP et al. (1989) 'Localisation of endocrine-related tumours with radioiodinated analogue of somatostatin.' *Lancet* i: 242–244.

16. Scintigraphic procedures for the proof of peritoneo-venous shunt patency

WERNER WATERS

Summary

Peritoneous shunt patency can be evaluated scintigraphically after intraperitoneal injection of ^{99m}Tc-labelled macroaggregated albumin. Rapid accumulation of the particles in the lungs is an indicator of patency. If no radioactivity occurs in the lungs, the system is obstructed. After a second injection into the valve body the scintigraphic localization of the obstruction is possible.

Introduction

LeVeen and coworkers developed in 1974 a tube system for ascites therapy which pumps the ascites fluid continuously into the blood pool [3]. The LeVeen shunt consists of a silicon and propylen tube system and a pressure regulated one way valve. The perforated input tube is implanted in the abdomen. The output tube is introduced in the vena jugularis, with its tip in the vena cava superior. The valve is implanted subcutaneously, the output tube under the skin, as well. If the intra-abdominal pressure exceeds that of the vena cava superior by more than 3 to 5 cm H_2O, ascites will flow.

The function of the Denver shunt is similar to that of the LeVeen shunt. It was originally used for hydrocephalus therapy [2]. The essential difference is the simpler construction of the one way valve. Due to the elasticity of the valve's body it may be mechanically compressed from outside so that the ascites flow can be forced by manipulation [5].

The output piece can be introduced from the vena saphena magna into the vena cava, too [10]. Its tip must lie in the thorax, as during each inspiration the pressure gradient rises thus leading to a breathing assisted pump function.

Retrograde flow can cause fibrin adhesions in the valve, or blood clots in the tubes, that may lead to shunt obstruction. Numerous techniques have been developed for testing shunt patency. The most widespread technique

H.J. Biersack and P.H. Cox (eds), Nuclear Medicine in Gasteroenterology, 239–244

is the scintigraphic evaluation of the system using macroaggregated ^{99m}Tc-albumin [8].

Method

74 MBq (2 mCi) macroaggregated ^{99m}Tc-albumin are injected intraperitoneally. The radiopharmacon can be injected into the valve body, if occlusion shall be localized.

If the shunt is patent radioactivity is rapidly transported into the pulmonary circulation, the albumin particles being retained in the capillary system of the lung thus enabling its scintigraphic detection as usually performed in the perfusion scintigraphy of the lung. In most cases the output tube of the shunt system is seen, too. If complete imaging of the lungs is achieved after intraperitoneal injection within 5 min, no further scintigraphic acquisitions are necessary. If none or only minute amounts of radioactivity can be detected over the lung after 5 min, further scintigraphic acquisitions are needed, e.g., after 15 and 30 min. If no albumin particles are fixed in the lung after 1 h, complete occlusion has occured.

Then a second injection into the valve body is indicated in order to localize the occlusion. If the efferent tube is occluded it will be seen in the scintigram. Manipulation of the valve body leads to accumulation of radioactivity in the tube if the obstruction is not complete. Imaging of the tube as well as the lung will be achieved if the valve input is occluded.

Examples

Example 1

In a 55 years old male patient with ascites the radioactive albumin particles remained completely in the abdomen 3 h after intraperitoneal injection. The lungs were not visible (Fig. 1).

After the second injection into the valve the particles migrated into the efferent tube system. The obstruction was scintigraphically detected at the tip of the upper tube. Only minute amounts of radioactivity had reached the lungs (Fig. 2).

Example 2

In a 51 years old male patient with ascites radioactivity stayed in the abdomen even 5 h after the injection. The Denver-shunt was completely occluded. The overranged scintigram showed only lymph nodes of the mammariae internae chains as well as the thyroid and the salivary glands due to free ^{99m}Tc-pertechnetate (Fig. 3).

After surgical revision of the shunt, ascites occured occasionally in this patient. As the patient had at the time of reinvestigation no ascites, the

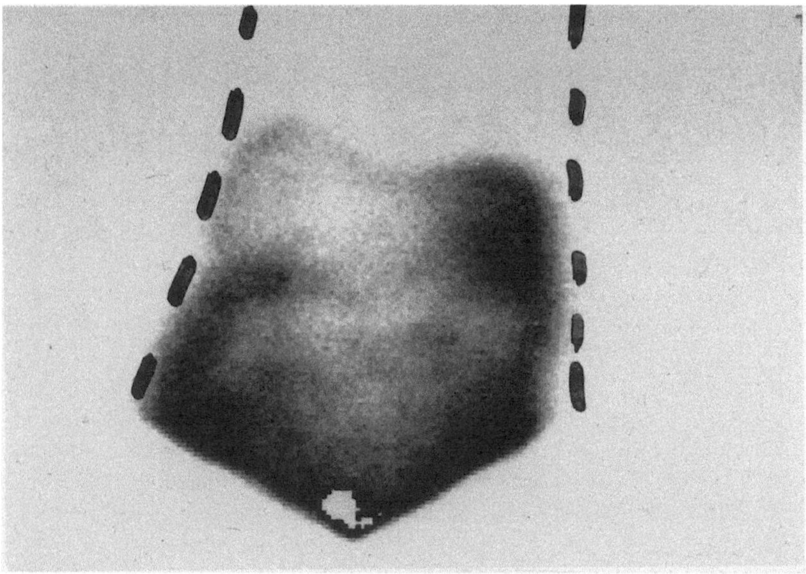

Figure 1. Complete obstruction of a LeVeen shunt. The radioactive macroaggregated albumin particles remain in the abdomen after intraperitoneal injection.

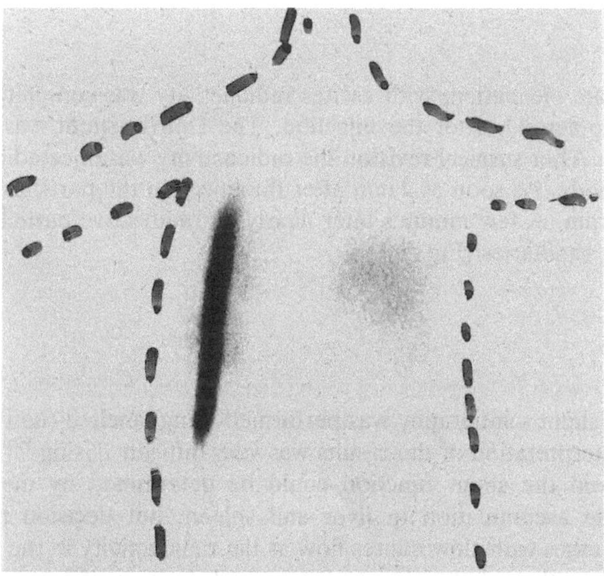

Figure 2. After the injection of the particles into the valve body of the shunt system the obstruction can be localized at the tip of the efferent tube.

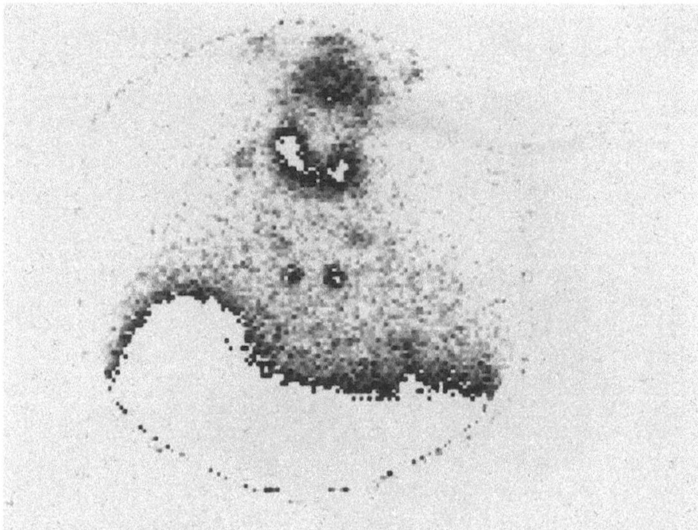

Figure 3. Complete obstruction of a Denver shunt. The radioactive particles remain in the abdomen. In the overranged scintigram only lymph nodes of the mammariae internae chains are seen, as well as the thyroid and the salivary glands due to free ^{99m}Tc-pertechnetate.

radiopharmacon was injected directly into the valve body of the shunt system. The particles were transported immediately into the lungs, so that the tube system was not seen (Fig. 4).

Example 3
In a 52 years old patient with ascites radioactivity was completely retained in the abdomen 4 h after the injection. The Denver-shunt was completely obstructed. After surgical revision the radioactivity was injected directly into the valve body. As soon as 2 min after the injection the particles are seen in the scintigram. A few minutes later nearly all radioactive particles are fixed in the lung capillaries (Fig. 5).

Discussion

Previously shunt scintigraphy was performed using labelled (native) albumin [6]. The interpretation of the results was very difficult. Using ^{99m}Tc-labelled sulfur colloid the shunt function could be determined by measuring the radionuclide accumulation in liver and spleen, but decision making was difficult in cases with slow ascites flow as the radioactivity in the ascites fluid was added to that of the target organs; and, in case of liver cirrhosis the colloid accumulation was diminished causing additional difficulties.

Nowadays we use ^{99m}Tc-labelled macroaggregated albumin particles, as

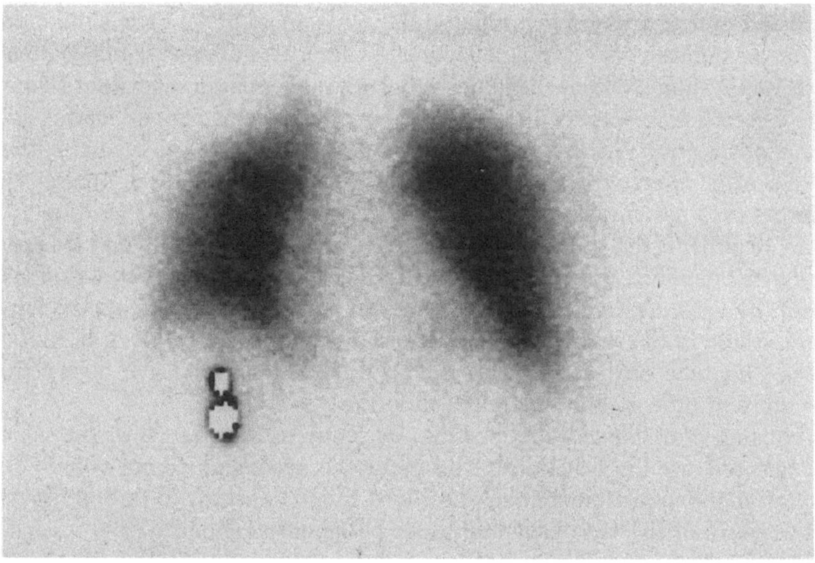

Figure 4. The same patient as in Fig. 3 after surgical revision of the shunt. Radioactivity in the lungs indicate that the shunt is patent.

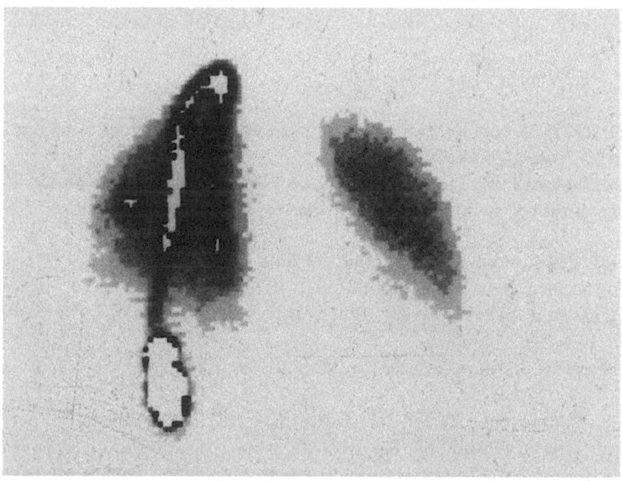

Figure 5. Patent Denver shunt with scintigraphic imaging of the efferent tube. Nearly complete particle fixation in the lungs 10 min after the intraperitoneal injection.

the target organ, the lung, is imaged without background radioactivity. The radioactivity is injected intraperitoneally or into the valve body. In the latter case ultrasound investigation may be very helpful defining the best injection site.

Stuart and coworkers investigated the function of 61 LeVeen shunts and 5 Denver shunts with ^{99m}Tc-labelled macroaggregated albumin. In 31 patients with patent shunt systems the lungs were seen in the scintigram within 10 min. The efferent tube was seen in 26 of 31 patent shunts. 4 patients had partial shunt obstruction. In these 4 patients the lungs were seen, but later than 10 min after injection (30, 35, 60 or 300 min). In two of these 4 cases the efferent tube system was scintigraphically seen.

In 30 patients the shunt was completely occluded, caused by tube obstruction in 19 patients, by wrong position in 3 patients (e.g., by bending of the tube), by valve obstruction in 8 patients. In none of these patients the lungs were visible in the scintigram. The efferent tube system was seen in 3 of 19 cases with tube obstruction. In cases of tube bending or valve obstruction the efferent tube system was not seen in the scintigram.

Imaging of a shunt is also possible by X-ray using contrast media which are injected into the system [7]. This method is accompanied not only by the risk of allergic reactions but can also lead to detatchment of thrombi in the tube system or the vena cava and cause pulmonary embolism [9].

Ascites flow can be measured by Doppler ultrasound [4], but the localization of an obstruction is not possible.

In the case of complete or incomplete shunt obstruction surgical revision is indicated. In order to optimize the operation the site of obstruction should be determined; in many cases then only part of the system must be revised.

References

1. Gorten RJ (1977) 'A test for evaluation of peritoneo-venous-shunt function: concise communication.' *J Nucl Med* 18: 29–31.
2. Kirsch WM, Newkirk JB, Predecki PK (1970) 'Clinical experience with the Denver-shunt: A new silicone-rubber shunting device for the treatment of hydrocephalus. *J Neurosurg* 32: 258–264.
3. LeVeen HH, Christoudias G, Ip M, Luft R, Falk G, Grosberg S (1974) 'Peritoneo-venous shunting for ascites.' *Ann Surg* 180: 580–591.
4. Metzler M, Lichti E, Silver D (1980) 'Noninvasive determination of LeVeen-shunt patency.' *Surgery* 87: 106–108.
5. Oosterlee J (1980) 'Peritoneovenous shunting for ascites in cancer patients.' *Br J Surg* 67: 663–666.
6. Rikkers LF, Fajman WA, Ansley JD, Tarcan YA (1977) 'Patency of the peritoneovenous shunt.' *Surg Gynecol Obstet* 145: 745–747.
7. Ring EJ, Rosato EF (1979) 'Assessment of peritoneojugular shunts by direct roentgenographic examination.' *Surg Gynecol Obstet* 148: 93–94.
8. Sakimura IT, Shapero T, Redeker A, Siemsen JK (1978) 'Intraperitoneal ^{99m}Tc-MAA for evaluation of peritoneovenous shunt patency (abstract).' Annual Western Regional Meeting of the Society of Nuclear Medicine, Vancouver BC.
9. Stewart CA, Sakimura IT, Applebaum DM, Siegel ME (1986) 'Evaluation of peritoneovenous shunt patency by intraperitoneal ^{99m}Tc-macroaggregated albumin: Clinical experience.' *AJR* 147: 177–180.
10. Turner WW, Pate RM (1982) 'The Denver peritoneovenous shunt: Relationship between hepatic reserve and successful treatment of ascites.' *Ann J Surg* 144: 619–623.

Index

Developments in Nuclear Medicine

1. P.H. Cox (ed.): *Cholescintigraphy*. 1981 ISBN 90-247-2524-0
2. P.H. Cox (ed.): *Progress in Radiopharmacology*. Selected Topics. Proceedings of the 3rd European Symposium (Noordwijkerhout, The Netherlands, April 1982). 1982 ISBN 90-247-2768-5
3. M.H. Jonckheer and F. Deconinck (eds.): *X-Ray Fluorescent Scanning of the Thyroid*. 1983 ISBN 0-89838-561-X
4. K. Kristensen and E. Nørbygaard (eds.): *Safety and Efficacy of Radiopharmaceuticals*. 1984 ISBN 0-89838-609-8
5. A. Bossuyt and F. Deconinck: *Amplitude/Phase Patterns in Dynamic Scintigraphic Imaging*. With a Foreword by A. Bertrand Brill. 1984 ISBN 0-89838-641-1
6. M.R. Hardeman and Y. Najean (eds.): *Blood Cells in Nuclear Medicine, Part I*. Cell Kinetics and Bio-distribution. 1984 ISBN 0-89838-653-5
7. G.F. Fueger (ed.): *Blood Cells in Nuclear Medicine, Part II*. Migratory Blood Cells. 1984 ISBN 0-89838-654-3
8. H.J. Biersack and P.H. Cox (eds.): *Radioisotope Studies in Cardiology*. 1985 ISBN 0-89838-733-7
9. P.H. Cox, G. Limouris and M.G. Woldring (eds.): *Progress in Radiopharmacology 1985*. 1985 ISBN 0-89838-745-0
10. P.H. Cox, S.J. Mather, C.B. Sampson and C.R. Lazarus (eds.): *Progress in Radiopharmacy*. 1986 ISBN 0-89838-823-6
11. H. Deckart and P.H. Cox (eds.): *Principles of Radiopharmacology*. 1987 ISBN 0-89838-774-4
12. W.-D. Heiss, G. Pawlik, K. Herholz and K. Wienhard (eds.): *Clinical Efficacy of Positron Emission Tomography*. 1987 ISBN 0-89838-898-8
13. G.B. Gerber, H. Métivier and H. Smith (eds.): *Age-related Factors in Radionuclide Metabolism and Dosimetry*. 1987 ISBN 0-89838-953-4
14. K. Kristensen and E. Nørbygaard (eds.): *Safety and Efficacy of Radiopharmaceuticals 1987*. 1987 ISBN 0-89838-986-0
15. C. Beckers, A. Goffinet and A. Bol (eds.): *Positron Emission Tomography in Clinical Research and Clinical Diagnosis*. Tracer Modelling and Radioreceptors. 1989 ISBN 0-7923-0254-0
16. M. De Schrijver: *Scintigraphy of Inflammation with Nanometer-sized Colloidal Tracers*. 1989 ISBN 0-7923-0272-9
17. Ch. Kessler, M.R. Hardeman, H. Henningsen and J.-N. Petrovici (eds.): *Clinical Application of Radiolabelled Platelets*. 1990 ISBN 0-7923-0729-1
18. H.J. Biersack and P.H. Cox (eds.): *Nuclear Medicine in Gastroenterology*. 1991 ISBN 0-7923-1074-8

Kluwer Academic Publishers - Dordrecht / Boston / London

MIX
Papier aus verantwortungsvollen Quellen
Paper from responsible sources
FSC® C105338

If you have any concerns about our products,
you can contact us on
ProductSafety@springernature.com

In case Publisher is established outside the EU,
the EU authorized representative is:
Springer Nature Customer Service Center GmbH
Europaplatz 3, 69115 Heidelberg, Germany

Printed by Libri Plureos GmbH
in Hamburg, Germany